# Communication Technology

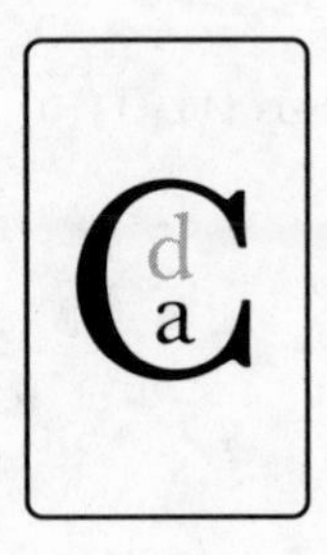

Published in association with the Centre for Canadian Studies at Mount Allison University. Information on the Canadian Democratic Audit project can be found at www.CanadianDemocraticAudit.ca.

**Advisory Group**

William Cross, Director (Mount Allison University)
R. Kenneth Carty (University of British Columbia)
Elisabeth Gidengil (McGill University)
Richard Sigurdson (University of Manitoba)
Frank Strain (Mount Allison University)
Michael Tucker (Mount Allison University)

**Titles**

John Courtney, *Elections*
William Cross, *Political Parties*
Elisabeth Gidengil, André Blais, Neil Nevitte, and Richard Nadeau, *Citizens*
Jennifer Smith, *Federalism*
Lisa Young and Joanna Everitt, *Advocacy Groups*
David Docherty, *Legislatures*
Graham White, *Cabinets and First Ministers*
Darin Barney, *Communication Technology*
Ian Greene, *The Courts*

# COMMUNICATION TECHNOLOGY

Darin Barney

15 14 13 12 11 10 09 08 07 06 05     5 4 3 2 1

Printed in Canada on acid-free paper that is 100% post-consumer recycled, processed chlorine-free, and printed with vegetable-based, low-VOC inks.

**Library and Archives Canada Cataloguing in Publication**

Barney, Darin David, 1966-
Communication technology / Darin Barney.

(Canadian democratic audit; 8)
Includes bibliographical references and index.
ISBN 0-7748-1101-3 (set). – ISBN 0-7748-1182-X (bound); ISBN 0-7748-1183-8 (pbk)

1. Democracy – Canada. 2. Information technology – Canada – Political aspects. 3. Telecommunication – Canada – Political aspects. 4. Political participation – Canada. I. Title. II. Series.

JL186.5.B37 2005     321.8'0971     C2005-901911-5

Canada

UBC Press gratefully acknowledges the financial support for our publishing program of the Government of Canada through the Book Publishing Industry Development Program (BPIDP), and of the Canada Council for the Arts and the British Columbia Arts Council.

The Centre for Canadian Studies thanks the Harold Crabtree Foundation for its support of the Canadian Democratic Audit project.

Printed and bound in Canada by Friesens
Copy editor: Sarah Wight
Text design: Peter Ross, Counterpunch
Typesetter: Artegraphica Design Co. Ltd.
Proofreader: Gail Copeland
Indexer: Aaron Gordon

UBC Press
The University of British Columbia
2029 West Mall
Vancouver, BC V6T 1Z2
604-822-5959 / Fax: 604-822-6083
**www.ubcpress.ca**

For Punker and Willie

# Contents

# Foreword

This volume is part of the Canadian Democratic Audit series. The objective of this series is to consider how well Canadian democracy is performing at the outset of the twenty-first century. In recent years, political and opinion leaders, government commissions, academics, citizen groups, and the popular press have all identified a "democratic deficit" and "democratic malaise" in Canada. These characterizations often are portrayed as the result of a substantial decline in Canadians' confidence in their democratic practices and institutions. Indeed, Canadians are voting in record low numbers, many are turning away from the traditional political institutions, and a large number are expressing declining confidence in both their elected politicians and the electoral process.

Nonetheless, Canadian democracy continues to be the envy of much of the rest of the world. Living in a relatively wealthy and peaceful society, Canadians hold regular elections in which millions cast ballots. These elections are largely fair, efficient, and orderly events. They routinely result in the selection of a government with no question about its legitimate right to govern. Developing democracies from around the globe continue to look to Canadian experts for guidance in establishing electoral practices and democratic institutions. Without a doubt, Canada is widely seen as a leading example of successful democratic practice.

Given these apparently competing views, the time is right for a comprehensive examination of the state of Canadian democracy. Our purposes are to conduct a systematic review of the operations of Canadian democracy, to listen to what others have to say about Canadian democracy, to assess its strengths and weaknesses, to consider where there are opportunities for advancement, and to evaluate popular reform proposals.

A democratic audit requires the setting of benchmarks for evaluation of the practices and institutions to be considered. This necessarily involves substantial consideration of the meaning of democracy.

"Democracy" is a contested term and we are not interested here in striking a definitive definition. Nor are we interested in a theoretical model applicable to all parts of the world. Rather, we are interested in identifying democratic benchmarks relevant to Canada in the twenty-first century. In selecting these we were guided by the issues raised in the current literature on Canadian democratic practice and by the concerns commonly raised by opinion leaders and found in public opinion data. We have settled on three benchmarks: public participation, inclusiveness, and responsiveness. We believe that any contemporary definition of Canadian democracy must include institutions and decision-making practices that are defined by public participation, that this participation must include all Canadians, and that government outcomes must respond to the views of Canadians.

While settling on these guiding principles, we have not imposed a strict set of democratic criteria on all of the evaluations that together constitute the Audit. Rather, our approach allows the auditors wide latitude in their evaluations. While all auditors keep the benchmarks of participation, inclusiveness, and responsiveness central to their examinations, each adds additional criteria of particular importance to the subject he or she is considering. We believe this approach of identifying unifying themes, while allowing for divergent perspectives, enhances the project by capturing the robustness of the debate surrounding democratic norms and practices.

We decided at the outset to cover substantial ground and to do so in a relatively short period. These two considerations, coupled with a desire to respond to the most commonly raised criticisms of the contemporary practice of Canadian democracy, result in a series that focuses on public institutions, electoral practices, and new phenomena that are likely to affect democratic life significantly. The series includes volumes that examine key public decision-making bodies: legislatures, the courts, and cabinets and government. The structures of our democratic system are considered in volumes devoted to questions of federalism and the electoral system. The ways in which citizens participate in electoral politics and policy making are a crucial component of the project, and thus we include studies of advocacy groups and political parties. The desire and capacity of Canadians for meaningful participation in public life is also the subject of a volume. Finally, the challenges and opportunities raised by new communication technologies are also considered. The Audit does not include studies devoted to the status of particular groups of Canadians. Rather than separate out Aboriginals, women, new Canadians, and others, these groups are treated together with all Canadians throughout the Audit.

In all, this series includes nine volumes examining specific areas of Canadian democratic life. A tenth, synthetic volume provides an overall assessment and makes sense out of the different approaches and findings found in the rest of the series. Our examination is not exhaustive. Canadian democracy is a vibrant force, the status of which can never be fully captured at one time. Nonetheless the areas we consider involve many of the pressing issues currently facing democracy in Canada. We do not expect to have the final word on this subject. Rather, we hope to encourage others to pursue similar avenues of inquiry.

A project of this scope cannot be accomplished without the support of many individuals. At the top of the list of those deserving credit are the members of the Canadian Democratic Audit team. From the very beginning, the Audit has been a team effort. This outstanding group of academics has spent many hours together, defining the scope of the project, prodding each other on questions of Canadian democracy, and most importantly, supporting one another throughout the endeavour, all with good humour. To Darin Barney, André Blais, Kenneth Carty, John Courtney, David Docherty, Joanna Everitt, Elisabeth Gidengil, Ian Greene, Richard Nadeau, Neil Nevitte, Richard Sigurdson, Jennifer Smith, Frank Strain, Michael Tucker, Graham White, and Lisa Young I am forever grateful.

The Centre for Canadian Studies at Mount Allison University has been my intellectual home for several years. The Centre, along with the Harold Crabtree Foundation, has provided the necessary funding and other assistance necessary to see this project through to fruition. At Mount Allison University, Peter Ennals provided important support to this project when others were skeptical; Wayne MacKay and Michael Fox have continued this support since their respective arrivals on campus; and Joanne Goodrich and Peter Loewen have provided important technical and administrative help.

The University of British Columbia Press, particularly its senior acquisitions editor, Emily Andrew, has been a partner in this project from the very beginning. Emily has been involved in every important decision and has done much to improve the result. Camilla Blakeley has overseen the copyediting and production process and in doing so has made these books better. Scores of Canadian and international political scientists have participated in the project as commentators at our public conferences, as critics at our private meetings, as providers of quiet advice, and as referees of the volumes. The list is too long to name them all, but David Cameron, Sid Noel, Leslie Seidle, Jim Bickerton, Alexandra Dobrowolsky, Livianna Tossutti, Janice Gross Stein, and Frances Abele all deserve special recognition for their contributions. We are also grateful to the Canadian Study of Parliament Group, which partnered with us for our inaugural conference in Ottawa in November 2001.

Finally, this series is dedicated to all of the men and women who contribute to the practice of Canadian democracy. Whether as active participants in parties, groups, courts, or legislatures, or in the media and the universities, without them Canadian democracy would not survive.

William Cross
Director, The Canadian Democratic Audit
Sackville, New Brunswick

## Acknowledgments

I am indebted to Bill Cross, for his leadership; to the members of the Canadian Democratic Audit team, for their intellectual generosity and scholarship; to Mary Stone for her heart; to Peter Milroy and Ken Carty, for their vision; to Emily Andrew, for her unique ability and endless patience; to Camilla Blakeley, for her appreciation of what is at stake in details; to Sarah Wight, for her fine-toothed comb; to Tim Plumbtre, Sid Noel, and all invited and occasional critics of the Audit's work-in-progress, for their insight and inquiries; to Peter Hodgins, for his friendship and curiosity; to Leslie Shade, Barbara Crow, Graham Longford, and all the other scholars of communication in Canada whose excellent work is mined in these pages; to Jennifer Hefler, for her detective work; to Aaron Gordon, for a whole list of things; to the anonymous reviewers of this manuscript, for their critical insight; and, finally, to Tad Beckman, Dick Olson, Pat Little, and Michael Black of Harvey Mudd College, where Chapters 2 and 3 were written, for teaching me the meaning of collegiality.

Communication Technology

# 1 DEMOCRACY, TECHNOLOGY, AND COMMUNICATION IN CANADA

The 2000 Canadian general election, understood at the time to be the country's first "Internet election," also featured the lowest voter turnout in the history of these contests at the federal level. Just 61 percent of registered voters turned out to cast ballots – when measured against the entire voting-age population, this figure drops to 55 percent (Johnston 2001, 13). Significantly, these numbers are less a blip than the continuation of a trend that has seen voter turnout in Canada drop precipitously and consistently from 75 percent of registered voters in the 1988 election to 71 percent in 1993, to 67 percent in 1997, and finally to the millennial level of 61 percent. This downward trajectory in this most basic form of political participation has occurred during the same period of time that formidable new information and communication technologies have come to occupy the Canadian political landscape and fairly saturate the Canadian political imagination. In its 1999 speech from the throne, the government of Canada articulated its goal "to be known around the world as the government most connected to its citizens" (Canada 1999); two years later it declared that it had "helped to make Canada one of the most connected countries in the world" (Canada 2001). This was no idle boast, as Canada does indeed rank highly among industrialized nations on most measures of Internet connectivity. It is also the case that, as political scientist Richard Johnston (2000, 13) has observed, recent electoral history "puts Canada near the bottom of the industrialized world turnout league tables ... No other G7 country besides the US has turnout as low as Canada's."

Admittedly, voter turnout is neither the only nor, arguably, even the best measure of the health of a democracy, and many factors combine to determine its level at any given time. The suggestion here is certainly not that the explosive growth of new information and communication technologies directly correlates with the decline in voter turnout in recent Canadian elections. That being said, the fact of their coincidence is provocative. One of our deepest liberal democratic intuitions is that generalized advance in our ability to gather and share information, and to communicate with one another, invigorates democratic participation. This intuition has received forceful expression in relation to the computerized and networked information and communication technologies (ICTs) that mediate an increasing array of social, political, and economic activity in Canada. Information and communication, we believe, are foundational to democracy, and therefore technologies that facilitate these contribute positively to democracy's achievement and enhancement. How could a technology such as the Internet, which provides widespread instant access to increasing volumes of politically relevant information, and which enables direct, undistorted communication among citizens (and rulers) be anything other than complementary to informed, democratic deliberation and self-government?

The coincidence of the rise of the Internet and a historic decline in voter turnout does not invalidate the hypothesis that ICTs will enhance democracy in Canada. It does, however, raise the possibility that recent technological advances in information and communication capacity are not unambiguously or automatically beneficial to Canadian democracy, nor capable of overcoming other factors that may contribute to its current condition. Indeed, one of the nasty little facts of the coincident growth of mass democracy and mass media in the twenty-first century is that despite a dramatic trajectory of technological expansion in information and communication capacity, democratic participation has not improved significantly in quantitative or qualitative terms. As Bruce Bimber has written, documenting the absence of statistical evidence linking Internet use to increased political engagement (in its various forms) in the United States:

> Opportunities to become better informed have apparently expanded historically, as the informational context of politics has grown richer and become better endowed with media and ready access to political communication. Yet none of the major developments in communication in the twentieth century produced any aggregate gain in citizen participation. Neither telephones, radio, nor television exerted a net positive effect on participation, despite the fact that they apparently reduced information costs and improved citizens' access to information (Bimber 2001, 57).

While we must be sensitive to the technical attributes that distinguish new from previous mass media, we must also acknowledge the ways in which they may be the same. Similarly, we must be as open to the possibility that politics mediated by new technologies will aggravate the disconnection between information/communication and democratic engagement as we are to our intuition that they will mediate a democratic renaissance.

This suggests that the relationship between ICTs and Canadian democracy is more of a problem to be explored than a foregone conclusion. It is a problem that exists at a very basic philosophical level, a problem that has manifested itself historically in Canada, and a problem that surfaces in particular ways in the contemporary context of new ICTs. For many reasons that will become evident through the course of this investigation, the problem of democracy, technology, and communication crystallizes broader dynamics and questions of democratic citizenship, identity, power, and the public good. In this sense, democratic questions about technology and communication are something of a crucible, especially in the Canadian context.

This exploration of the relationship between ICTs and democracy in Canada will be framed by the three criteria set out for the Canadian Democratic Audit: public participation, inclusiveness, and responsiveness. Public participation is the sine qua non of democratic politics and government. Though participation can take many forms and be enacted in a variety of venues, the degree to which citizens take part in various processes of political expression, decision making, and governance is an indispensable measure of democratic legitimacy. Participation is an important concept for assessing the politics of ICTs in several respects. Have political processes surrounding the development and regulation of these technologies been participatory or not? Do ICTs provide means for improving or expanding political participation in Canada? And do ICTs enhance, or undermine, the socioeconomic equality that supports effective political participation?

Inclusiveness is the second Audit criterion, and it too is related to the core democratic principle of equality. Exclusivity, or privilege, is anathema to a democracy, wherein political participation must be at least available to, and at best undertaken by, as many citizens as possible without prejudice. A political order that formally or practically excludes significant segments of its citizenry from effective participation will be far less democratic than one that provides for inclusion of as many people as possible in the political process. This criterion is especially important in Canada, whose population exhibits multiple diversities that often correspond to systemic forms of disadvantage and exclusion. Here again, special questions are raised about ICTs. Has decision making surrounding their development and regulation included the diversity of views and interests of relevant constituencies in Canada? Do ICTs provide a means of effectively including a greater diversity of Canadians in political life? And have these technologies contributed to, or undermined, the socioeconomic basis of inclusion and political equality in Canada?

The third Audit criterion is responsiveness. It measures the degree to which various elements of the political system actually address, and are affected by, the needs, priorities, and preferences expressed by citizens in their participatory activities. In democratic polities, a diverse range of citizens participate not simply to lend the appearance of legitimacy to processes that may not *really* take their views into account; they participate so that political outcomes will reflect, at least to some degree, their duly expressed interests. In representative systems such as Canada's, the responsiveness of political agencies and institutions is a crucial measure of the democratic acceptability of a given regime. As with the criteria of participation and inclusiveness, ICTs have a special bearing on the question of responsiveness, and vice-versa. Has the development of ICT policy in Canada been sufficiently responsive to the diversity of interests at stake in it? Has the relationship between ICTs and globalization enhanced or diminished the capacity of Canadian governments to be responsive to their citizens? And has the use of ICTs by a variety of political actors made Canadian political institutions more responsive to public participation?

Taken together, the three criteria of participation, inclusiveness, and responsiveness focus the investigation that follows on three central questions:

- To what extent has the development of digital communication technology in Canada been subjected to democratic political judgment and control?
- What effect is the increasing mediation of political communication by digital technologies having on the practices of democratic politics in Canada?
- How do digital technologies affect the distribution of power in Canadian society?

These questions derive from an understanding that communication technology occupies a complex position in the universe of Canadian democracy. Communication and its mediating technologies are at once an object and an instrument of democratic practice in Canada. They also affect the material context in which democratic politics and citizenship take place. To concentrate on one of these questions to the exclusion of the others would be to tell only part of the story. I will return to consider the rationale that supports these questions later in this chapter. At this point, some added reflection is in order on the conception of democracy driving this investigation.

## Democracy

Politics admits of many definitions, practices, and expressions. Nevertheless, at its core, politics involves collective judgment by citizens regarding common goods, and the engagement of authoritative collective action toward the realization of those goods. Insofar as it reflects this combination of judgment and action, the ultimate practice of politics is often specified as governing or government. (These terms are derived from the ancient Greek *kubernetes,* or "steersman," since to steer, one must form a judgment as to where the ship should go and take action to guide it there.) Politics, then, is not about strictly individual determinations of right and wrong conduct in personal affairs (the province of ethics), nor does it comprise simply those individual calculations of purely private self-interest that tend to guide economic behaviour in markets. Despite the many forms its constituent practices can take, genuine politics always has a public, collective character, it always involves judgment and action, and it always pursues goods identified as common.

Democracy is a particular manner of constituting the various practices of judgment and action that together make up politics. That is to say, democracy is a form of self-government. It casts the net of citizenship broadly, extending rights to participate in collective judgment (whether direct, delegated, or representative) on the basis of principles of equality, and deriving authority for sovereign acts from majoritarian consent. Within those parameters, existing democratic practices take many institutional and noninstitutional forms, which vary in the quality and degree of participation, deliberation, representation, inclusiveness, and legitimacy they embody. What unites these various practices as democratic is that each subjects matters pertaining to the common welfare to some manner of political judgment by citizens regarded as equals, and each maintains a discernible link between these judgments and the authoritative actions of government.

The stipulations set out above certainly allow for minimalist constructions of liberal democracy. For example, democracy can mean little more than periodic elections in which citizens who are formally equal express their private preferences by voting: a registration of consent that subsequently legitimizes the actions of a government. On its better days, however, democracy typically involves somewhat more. Even in representative democracies in which the main political activity for most citizens is voting in periodic elections, citizenship ought to exist as much between elections as it does during them, in the ongoing ability of people to contribute to common decision making in a meaningful manner. The word "meaningful" here means that, in a democracy, civic participation must be obviously connected to outcomes and it must be more than merely symbolic. Furthermore, even in liberal democracies that emphasize opportunity as the pivot upon which equality turns, there ought to be some recognition that not all people are equally *able* to take advantage of the citizenship opportunities afforded by their constitution. Thus, a robust democracy will seek out ways to equalize participatory ability so that it matches opportunity. Finally, while it is certainly possible for a democracy to serve as nothing more than a means of registering and aggregating private self-interest, a more substantial democracy will make the effort to orient its politics around civic deliberation on the common good, slippery though it may be. To adopt the language of one of democracy's great thinkers, Jean-Jacques Rousseau, democracy does not reside primarily in the combination of individual particular wills into the will-of-all, but rather in public-spirited generation of the general will.

Together, these stipulations give added substance to the criteria of participation, inclusiveness, and responsiveness used throughout this book. They construct an understanding of democracy that is neither radical nor foreign to the Canadian experience. Canadians understand their society to be democratic, and by that I think we can assume they mean more than that they get to vote occasionally. They probably mean that theirs is a political system in which *inclusiveness*, public *participation*, and *responsiveness* – the benchmarks of the Canadian Democratic Audit – are legitimate demands that citizens can reasonably expect will be met. This does not mean that Canadian democracy is perfectly or even sufficiently inclusive, participatory, and responsive. Rather, Canada is a democracy to the extent that serious deficits of inclusiveness, participation, and responsiveness are widely understood by its citizens to be illegitimate and intolerable. Far from containing a utopian standard that prejudicially disqualifies Canada as a democracy, this formulation arguably captures the kind of democracy Canada and Canadians imagine themselves as striving to be. The underlying question of this study is whether and to what extent our current encounter with ICTs contributes to meeting this goal.

These technologies, however, are not the only factor involved in securing a democratic political order on the terms outlined above. Indeed, the impact of ICTs on democracy can really be understood only in light of, or in relation to, a number of other conditions necessary to sustain a democracy. As I will discuss in further detail in Chapter 5, these conditions include not just a democratic constitution that distributes effective political power equally, but also an economy in which the material resources crucial to citizenship are distributed relatively equally, a culture in which the habits of citizenship are the norm rather than the exception, and a public sphere in which politics are conspicuous by their presence, rather than by their absence. Inclusive, participatory, responsive democracies require all of these conditions, whether or not technology is part of the picture. As I will argue in Chapter 5, however, when technology *is* part of the picture it has a significant impact upon the possibility of these conditions being met, and this has been especially true of ICTs in the contemporary period.

## Technology

Canada is not only a democratic society. It is also an unambiguously technological society. Since at least the Second World War, Canada's commitment to democratic politics has been matched by a resolute commitment to the development of technology as a means to secure its material economic well-being. Statements from the government of Canada regarding "the challenge and the urgency" of constructing the "Information Highway" are but the latest manifestation of this enduring technological conviction (Industry Canada 1996, 3). But Canada's democratic convictions may be at odds with its technological commitments on a fundamental level, as a technological society may not be able to either support or withstand the sort of decision making and action described above as democratic, and still remain a technological society.

The tension between technology and democracy has three aspects. The first is that the complexity of technological issues can undermine the possibility of either intensive or extensive democratic consideration. Democracies do not demand expertise of their citizens as a condition of participation, but technological complexity can make demands that exceed the capacities of most citizens, thus reducing the efficacy of citizenship.

Second, even if the majority of citizens had the capacity to engage with complicated technological issues, their deliberations would most certainly undermine the conditions in which technology develops and is optimized. Democratic decision making tends to be slow, ponderous, risk averse, prone to reversals, lacking in clarity, easily seduced by superficial imaginings, and often irrational: qualities inimical to technological enterprise. It might not be to the material advantage of a technological society to subject technical determinations to genuine democratic consideration on a routine basis.

Third, modern technology tends to be universal rather than local, a quality that has been raised to high relief by new ICTs and their relationship to the phenomenon known as globalization. Technologies, especially those whose operation transcends national boundaries, challenge the applicability and enforceability of democratic political decisions and actions organized at the national level. Canadians have experienced this problem for a long time, especially in regard to communication technologies and policies: technologies that tend to transcend constraints of territorial space as a matter of their very design versus policies that are confined in their application to the territory over which the Canadian state is sovereign. Put simply, the democratic political authority of the Canadian state over broadcasting stops at the country's southern border, but radio signals originating from south of that border know no such constraint. Similarly, with regard to a technology such as the Internet, it could be argued that the wishes of the Canadian state – democratically derived or otherwise – are irrelevant to the terms under which this technology will be developed as a global phenomenon, and that Canada's only choice is whether or not it wishes to be part of the world connected by this technology. In this case, for a society committed to technological development as a condition of its material progress, the choice is self-evident.

This suggests that a society that imagines itself as democratic has to be willing to pay the price of restraint, regression, and inefficiency in technological matters. It also raises the possibility that a society devoted to technological progress as a condition of its material prosperity may not be able to maintain a commitment to democracy that is anything more than rhetorical when it comes to technological matters. Technology recommends technocracy – rule by experts – over democracy. And technological matters are regularly given over specifically to experts intimate with the imperatives of science, management, and the market, regimes whose ends and practices are rarely accused of being particularly democratic and which typically shield technological issues from potentially obtrusive democratic consideration. Precisely this tendency prompted Ursula Franklin (1999, 121) to observe, radically, that in Canada "we now have nothing but a bunch of managers who run the country to make it safe for technology." There seems to be something deeply depoliticizing and fundamentally undemocratic about technology.

But that is not the whole story. Although democratic political deliberation can sometimes slow down technological innovation, technology is also irreducibly political. Far from being mere instruments or tools that accomplish their direct ends and nothing else, technologies also condition priorities, define possibilities, set limits on practices, constitute infrastructures and environments, and mediate relationships between individuals and between people and the natural world. As the American political theorist Langdon Winner (1995, 67) has written, when it comes to technology "the central issues concern how the members of society manage their common affairs and seek the common good. Because technological things so often become central features in widely shared arrangements and conditions of life in contemporary society, there is an urgent need to think of them in a political light." In a similar vein, Franklin (1999, 120) characterizes questions concerning technology specifically as questions of "governance." For example, grain elevators are not simply instruments for handling grain. They also organize communities economically and spatially, and provide the material infrastructure for an entire way of living on the Canadian prairies. Their "progressive" replacement by high-throughput grain terminals is, consequently, radically restructuring communities and ways of living that grew up around the previous technology. The decision to replace the old elevators with the new terminals did not clearly emerge from an inclusive democratic political process that genuinely engaged and responded to the participation of those citizens whose lives are most affected by this technological change. Nevertheless, a technological moment such as this cannot be said to be without politics simply because its political aspect has been obscured by a perceived technological imperative. Technologies, in this sense, have a legislative character, insofar as they enable or encourage certain common practices and prohibit or discourage others. Technologies represent decisions about how we will and will not live together. Therefore no satisfactory meaning of the word "political" can exclude technologies and their effects.

Thus, technologies are political because they constitute widely shared social arrangements that frame a broad range of human social, political, and economic priorities and practices, and because they are artifacts in which power is embedded and through which power is exercised. Consequently, moments of technological change especially have the potential to be moments of intense democratic political contest, moments of deliberation over the character and needs of the common interest relative to the technology in question. These moments can also be sacrificed to the logic of depoliticization that is embedded in the technological spirit, which is often called forth by those who stand to benefit from insulating issues of technology from democratic political scrutiny. The history of the deployment of technologies of mass communication in Canada, and policy making surrounding this deployment, is replete with examples of this dynamic.

The political questions surrounding communication technology and policy in Canada have remained relatively consistent since at least the advent of the telegraph. They include questions about the following:

- the role of the state relative to the market in the distribution of communication resources
- the priority of either national-cultural or commercial-industrial objectives, and the tension between them
- the democratic imperative to ensure universal access to communication services throughout the country and the means to achieve it
- the liberal imperative of free expression in communication
- the structure of ownership and regulation in Canadian communication industries, including the possibility of state ownership
- the need to stimulate and secure domestic production and consumption of cultural content
- the role of public consultation in communication policy making
- the importance of separating control over carriage infrastructure (i.e., the pipes) from control over content (i.e., what goes through the pipes).

What is interesting about these enduring questions of Canadian communication policy historically is that, just as they begin to reach a point of settlement in relation to one communication medium, a technological change reopens them. Just when the politics surrounding the telegraph, for example, appeared to subside into normalization, the advent of the telephone repoliticized all the same old questions. It is also interesting to note the historical regularity with which technologically determinist arguments and rhetoric surface during times of technological change in communication – arguments and rhetoric often aimed at obscuring and depoliticizing the deeply political and highly contingent character of policy in this area. This strategy extrapolates from particular characteristics of the technology to specific policy choices that are presented as necessary outgrowths of the technology itself and, therefore, non-negotiable. This tactic is most often employed by those interests that have a great deal to gain in a particular configuration of technological change and a great deal to lose in political, and especially democratic, consideration of possible options.

A stark example is the "systems integrity" arguments used by telephone companies in the early and middle decades of the twentieth century to justify structuring the telephone industry in Canada as a natural monopoly. They argued that the technology involved in the successful construction and maintenance of a high-quality telephony system simply required that the system be controlled from end to end by a single entity, and ruled out other options from political consideration. The degree to which this technologically determinist argument became policy orthodoxy is suggested in *Instant world,* the 1971 report of a federal task force on telecommunications, which conceded that telephone companies had presented "powerful technical arguments for complete control of the public networks, including terminal devices and attached equipment. To maintain a high quality of service to all users, they must be able to guard against the technical pollution of the network from other signal sources" (DOC 1971, 156).

As we will see, there is no shortage of contemporary claims regarding the necessary connection between various technical aspects of new digital information and communication technologies and particular policy outcomes that are presented as non-negotiable. Interestingly, many of these – such as the suggestion that the technical properties of digital communication technologies demand competition and minimal regulation if they are to develop to their fullest potential in Canada – contradict the substance of earlier technologically determinist arguments such as those entailed in the "systems integrity/natural monopoly" thesis. This would seem to indicate that such arguments at times of technological change are themselves deeply strategic and political, and that the extent to which they are accepted by policy makers reflects the distribution of political power in Canada more than it does any inherent technological necessity. Curiously, the surfacing of technologically determinist rhetoric in moments of technological change can be read as evidence of the essentially political character of those moments.

This is not to say that technologies do not constrain and condition political options. A strong tradition in the philosophy of technology – to which Canadians such as George Grant (1998) have made enduring contributions – asserts that something in the essence of technology prescribes a particular way of being in the world, a particular way of relating to our environment and to those with whom we share it, to the exclusion of other ways. In a society where technology is ubiquitous and technological progress is an overwhelming collective social project, certain ways of living recommend themselves, persuasively, at the expense of others. It is to this quality that Canadian political economist and theorist of communication Harold Innis (1995, xxvii) referred when he suggested that communication technologies do more than enable us to communicate, and emphasized "the importance of communication in determining 'things to which we attend.'" Innis's concern was primarily with how *all* communication technologies reorient the human experience of space and time, and consequently reorganize human priorities and practices. Different communication technologies accomplish this in different ways, but the fact that each of them alters our natural experience of space and time can be said to belong to their essence as technologies. A great deal distinguishes a telephone from an automobile and both of these from a pile-driver; indeed, among communication technologies alone, a great deal distinguishes a telephone from a radio and both of these from the Internet. Still, despite these distinctions, all of these devices share a quality as technologies, a quality that makes the world we inhabit a technological one that is very different than a nontechnological world might be (assuming that we can conceive of such a thing). One need only try to imagine what life would be like in a world *without* technology to appreciate that there is something about technology in general that, despite the specificities of particular technological instruments, shapes our world, our practices, and our attention.

This observation returns us to the tension between technology and politics in general, and between technology and democratic politics in particular. On the one hand I have argued that technologies, and moments of technological change, are deeply political and open to contestation. On the other hand, I have suggested that technology in general, and specific technologies in particular, have essential characteristics that act to condition and limit available political options. Can both of these claims be true? Part of the answer lies in recognizing that a number of elements combine to produce any technological outcome or effect, and that varying degrees of political intervention are possible relative to these elements. Certainly, that which belongs to the essence of technology does not readily admit of political intervention, democratic or otherwise. But the practical outcome of a specific technology in the world is not wholly determined by its essence as a technology. A host of other factors – including design, situation, and use – also contribute to specific technological outcomes, and these typically exhibit considerable contingency, potentially leaving room for political determination.

Design refers to the technical configuration and orientation of a device's operation and application. Technological instruments are designed to do certain things in certain ways, and design choices can have serious political consequences. Referring specifically to the evolving design of the Internet and related technologies, American legal scholar Lawrence Lessig (1999, x, 3) has written that code builds "architectures of control" and so "code is law." In this sense, the effects of design are always political, and so too are the choices that precede design decisions, whether those privileged to make them recognize their political character or not.

Appreciation of the politics inherent in technological design immediately raises the question of democracy: if decisions about the design of technologies are political, then, in a democratic society, should they not be subjected to democratic deliberation? The answer is yes, but as Andrew Feenberg relates in the following passage, democratic participation at the fundamental level of design is far from the norm in modern technological societies:

> Technology is power in modern societies, a greater power in many domains than the political system itself. The masters of technical systems, corporate and military leaders, physicians and engineers, have far more control over patterns of urban growth, the design of dwellings and transportation systems, the selection of innovations, our experience as employees, patients and consumers, than all the electoral institutions of our society put together. But, if this is true, technology should be considered as a new kind of legislation, not so very different from other public decisions. The technical codes that shape our lives reflect particular social interests to which we have delegated the power to decide where and how we live, what kinds of food we eat, how we communicate, are entertained, healed and so on ... But if technology is so powerful, why don't we apply the same democratic standards to it we apply to other political institutions? By those standards, the design process as it now exists is clearly illegitimate (Feenberg 1999, 131).

The democratic imperative attached to matters of technological design, and the failure of modern technological societies to observe that imperative, could not be expressed with greater force or clarity. Democratic intervention in technological design typically conjures images of excessive bureaucracy, inefficiency, and irrationality, each of which is presented as anathema to effective design and technological innovation. Democratic engagement with issues of technological design does not necessarily have to embody these negative qualities. Yet such charges have been used quite effectively to exclude citizens from participation in technological decision making, other than as isolated consumers choosing to buy or sell long after crucial design decisions have already been made. Canada's experience with the development of digital ICTs has been no exception in this regard. Questions of design have been readily referred to the expertise of scientists, engineers, and corporate executives, and evidence of democratic participation, inclusiveness, and responsiveness is conspicuous by its absence. Were the democratic audit of new ICTs in Canada confined to the matter of their design, its findings would be brief and unequivocally damning.

That being said, technological outcomes are not wholly determined by design. All technological instruments and practices are situated in complex social, political, and economic environments that strongly condition their possible elaborations in human practice. Whether the outcome of our encounter with ICTs is substantially democratic or not will depend to a great degree upon its social, political, and economic context. Of course, a great deal of contingency is at play in this respect. A democratic audit of these technologies and their prospects has to take these material conditions and contingencies into account, so much of the analysis that follows will be devoted in one way or another to this task. Subsequent chapters will, for example, pay close attention to the policy framework and economic conditions under which these technologies have been developed in the Canadian context, in an attempt to locate evidence of democratic success or failure.

Finally, a substantial portion of any technological outcome is constructed socially through the actual everyday uses to which institutions and people put a given technological device. The essence of technology challenges, but does not negate, human freedom; technological design favours, but does not determine, potential applications; and material situation conditions, but does not enforce absolutely, possible elaborations. Even George Grant (1986, 21), who clearly prioritizes the essential elements of technology, concedes that "the computer does not impose upon us the ways in which it will be used." Technological outcomes are linked crucially to use, and use admits a significant range of possibilities, many of which were not contemplated by design and some of which involve "democratic rationalizations" of technologies that were not intended for democratic use, or which are situated in conditions that are not otherwise democratic (Feenberg 1999, 12). Consequently, a democratic audit of new ICTs must also attend to the manner in which these instruments are actually used by political actors and institutions in the Canadian context, in order to determine whether these uses either reflect or encourage a democratic practice that is more, or less, participatory, responsive, and inclusive.

## Communication and Democracy in Canada

In specifying the distinctly political nature of human beings, Aristotle singled out our capacity to communicate. A human being, he argued, "is by nature a political animal" precisely because human beings are unique in their capacity to communicate with each other about common issues "of good and evil, the just and the unjust" (Aristotle 1995, 1253a2-7). Politics, especially democratic politics, is impossible without communication. Deliberating citizens share information and communicate their opinions and reasons with one another; citizens communicate with elected and appointed representatives who, in turn, communicate with constituents; governing authorities, whether administrative or legislative, solicit information from subjects and communicate with them in various forms of service and command.

In mass societies, the bulk of significant political communication is mediated by technology. It is not just that democratic politics cannot exist without communication: contemporary democracies such as Canada could not function without communication technologies. They play an indispensable role in advanced political systems analogous to the role of transportation technologies such as railroads, highways, and airplanes in advanced economies. For this reason, the stakes in issues surrounding these technologies are very high. We all know how intense the politics of roads and railways can be, and indeed have been in Canadian history, a fact that attests to the centrality of these technologies to economic life. The centrality of ICTs to democratic political life has generated a similar history of intensive political contestation in Canada. To raise but one example, the history of state broadcasting in Canada cannot be understood outside its origins in an epic political confrontation between the Canadian Radio League and the Canadian Association of Broadcasters, one result of which was the firm establishment of mass communication as a public interest issue in Canada (Raboy 1990, 17-47). The collective amnesia that typically accompanies moments of technological change notwithstanding, contemporary debates surrounding the development and character of new ICTs are best understood as a continuation of this history of politicization.

As suggested earlier in this chapter, ICTs have a complex relationship with democratic politics in Canada. In the first place, these technologies serve as a crucial infrastructure for an increasing array of political activities in Canada. This fact requires that an audit of the democratic character of this new environment of political communication attend to the question of whether these technologies are, or are likely to be, successful in mediating democratic politics according to some of the standards set out in the foregoing discussion. That is to say, we must investigate *the effect that increasing mediation of political communication by digital technologies is having on the practices of democratic politics in Canada,* including the practices of government, political parties, and citizens.

Second, these technologies also play an increasingly central role in the social, political, and economic lives of Canadians – our shared arrangements for living together – even for those who opt out of using them routinely. In one way or another, we all live in the world as it is built by and around new ICTs. Therefore we must inquire into *the manner in which these technologies affect the distribution of power in Canada.* We must also understand the elaboration of these technologies as itself a public issue of the highest significance, and recognize that a society that fails to subject this matter to adequate democratic consideration undermines its own claims to being a democracy. As such, we must also inquire into *the extent to which the rapid and massive development of digital information and communication technology in Canada has been subjected to democratic judgment and control.* Together, these inquiries yield a provisional conclusion as to the inclusive, participatory, and responsive nature of this aspect of contemporary Canadian democracy.

It is tempting to begin this investigation with the obvious question of how Canadian citizens and institutions are using ICTs in their political activities. The meaning and significance of these activities, however, can be understood only in the context of how ICTs have been treated as an object of citizenship, and the role they have played in restructuring the political possibilities of the Canadian state. So discussion of the political uses of ICTs will be deferred until Chapter 4. Chapter 2 assesses the democratic character of recent policy making surrounding new information and communication technologies in Canada. The aim here is to assess the participatory, inclusive, and responsive qualities of policy making in this field. Chapter 3 examines the relationship between new ICTs and national culture and sovereignty in Canada. Issues of technology, culture, international capital, and national sovereignty walk hand in hand through the history of communication policy making, scholarship, and activism in Canada. These issues have gained prominence once again in light of the intimate relationship between digital communication technology and globalization. The question addressed in this chapter is whether these dynamics bode well, or ill, for the prospect of an inclusive, participatory, and responsive Canadian democracy.

This provides important context for Chapter 4, which examines the uses to which ICTs have been put by democratic actors in Canada, with specific focus on government, political parties, advocacy groups and social movements, and citizens. Here the intent is to gauge whether digital mediation enhances or undermines the practice of democratic citizenship, according to the criteria of participation, inclusiveness, and responsiveness. Under the heading "Digital Divides," Chapter 5 also provides context for the prospects of democratic uses of ICTs, by examining the role these technologies have played in establishing the material setting in which democratic citizenship might be practised. Specific attention will be paid here to the relationship between ICTs and the distribution of power in Canada, and to the possibility of the latter's democratization. This chapter examines the digital divide in Canada, the political economy of ICTs, and the role of these technologies in the democratic public sphere. Chapter 6 offers some concluding reflections on the central themes of this portion of the Canadian Democratic Audit.

# 2 THE POLITICS OF COMMUNICATION TECHNOLOGY IN CANADA

On 5 September 2001, Jean Chrétien, the prime minister of Canada, played a round of golf with Tiger Woods, the best golfer in the world, in the pro-am section of the Canadian Open. The tournament entry fee was waived for the prime minister, but the $13,000 donation to charity typically required to duff around the links with Woods was not. As it turns out, the donation was paid on the prime minister's behalf by Jean Monty, the chairman and CEO of BCE Incorporated, the corporate sponsor of the Canadian Open. Along with sponsoring sporting events, BCE owns and operates the following, among other enterprises:

- Bell Canada and a string of regional telephone companies
- the Bell ExpressVu satellite television service
- the Telesat Canada satellite system
- the Bell Mobility cellular network
- Bell Intrigna communication services
- the CTV television network and its subsidiaries
- TQS broadcasting
- several specialty cable channels, including TSN and Discovery
- the *Globe and Mail* newspaper, including *Report on Business* and on-line operations
- the Sympatico/Lycos Internet portal and service provider.

In short, BCE is a telecommunications giant whose holdings straddle multiple media platforms and bridge the divide between content and carriage, thus controlling a substantial portion of the technological conduit for communication in Canada, as well as a great deal of what passes through it. After his round of golf, the prime minister cheerfully donned a Bell Canada baseball cap.

BCE and similar conglomerates are what they are, and they do what they do, because the sovereign government of Canada allows them to be and to do those things, as a matter of policy. In Canada, as in most other countries, the business of communication, in its various forms, is subject to regulation. As a condition of being in the business of communication, firms must observe certain rules and adhere to certain standards. These rules and standards take a variety of forms and address a variety of practices, but they are all formed and enforced by state authority. They all express decisions taken by government and its agencies as to what the public interest demands or, at least, which set of private interests should be served and which should be neglected when it comes to matters of communication. These decisions, which direct the regulation of communication practices and industries in Canada, are a reflection of public policy: the programmatic articulation of the purposes and priorities of public authority as it is exercised on behalf of the citizens of Canada.

Communication and its technologies – which constitute a massive industry vital to the health of the Canadian economy, a crucial engine and infrastructure for industrial and economic enterprise more generally, and an indispensable medium of cultural and political life – are among the most important public policy areas in Canadian society. Communication policy is thus something of a litmus test for democratic politics in Canada, in terms of both process and outcomes. If a democracy is a political system in which decisions about the most significant collective matters emerge from public processes that are adequately inclusive, participatory, and representative, then one basic measure of Canadian democracy is the extent to which recent communication policy making meets these standards. Furthermore, given the increasingly important role of communication and its technologies in

Canadian society and economy, it is also necessary to gauge whether the outcome of recent policy decisions in this area is likely to enhance, or undermine, the prospects of democratic control over these media.

It is from this perspective that the prime minister's BCE-sponsored round of golf with Tiger Woods is so very interesting. BCE Inc. dominates such a large portion of the Canadian mediascape because federal communication policy in Canada allows, or even encourages, it to do so. We might cite, in this regard, the decision by the Canadian Radio-television and Telecommunications Commission (CRTC) on 7 December 2000 to allow BCE Inc., Canada's largest telecommunications company, to acquire CTV and its subsidiaries, one of Canada's leading broadcasting companies (CRTC 2000). This clearly signalled the priority of current Canadian policy to permit unprecedented levels of consolidation, concentration, and cross-ownership in communication industries. This policy was evident also in subsequent decisions to allow the mergers of CanWest Global Communications with the Hollinger chain of newspapers, and the sale of the broadcaster TVA to Quebecor, which also owns the Sun chain of newspapers and the Videotron cable and Internet operation. We might ask whether, and to what extent, this policy priority – which clearly has broad implications for public life in Canada – was generated by a process that was adequately democratic. As one would expect, the voices of BCE and other industry heavyweights ring loud and clear in the arenas where communication policy is made in Canada, and this is true not just because the prime minister gets to golf with Tiger Woods, though the event does symbolize the intimacy between the Canadian state and the titans of the communication industry.

The questions are whether the voices of industry drown out those of ordinary Canadian citizens, whether the policy process is inclusive of a suitably broad representation of contending views, and whether the policy that emerges is as responsive to these civic perspectives as it is to the private interests of powerful economic actors in the communication industry. To ask these questions is to assert that, in a democracy, public policy ought to emerge from processes that are inclusive and participatory, and policy decisions ought to be responsive to those processes. This, arguably, is particularly true of policy involving communication technology, given its role in mediating social, political, and economic life more generally. This chapter will explore these questions as they have played out in public policy approaches, processes, and decisions surrounding the development of digital information and communication technologies in Canada. Such an examination requires us to view recent developments in light of the history, priorities, and norms of communication policy making in Canada, as it has unfolded over the past century.

## Communication Policy in Canada

Communication policy in Canada is predominantly a federal matter that encompasses a broad range of activities including broadcasting, telecommunications, publishing, and a variety of cultural industries. Digital ICTs touch all of these fields, but policy for the digital age has unfolded primarily under the rubric of telecommunications and broadcasting. Policy in this area centres on the Telecommunications Act (1993) and the Broadcasting Act (1991), and involves a variety of federal ministries, agencies, and programs. Among these, the CRTC, Industry Canada, and the Department of Canadian Heritage currently play leading roles.

Historically, the objectives of communication policy in Canada have been remarkably consistent:

- maintenance of a prosperous domestic communication industry, as well as a communication and technological infrastructure capable of serving the interests of the Canadian economy more generally
- protection and promotion of Canadian national culture and identity in the domestic media environment, including support for bilingualism

- adherence to the principles of democratic communication and communication policy making
- encouragement of technological innovation and progress as an engine of economic growth, material prosperity, and political independence.

References to these objectives, in one combination or another and with varying degrees of emphasis, appear with predictable regularity in the historical and contemporary documents of Canadian communication policy.

For example, Section 7 of the Telecommunications Act (1993) invokes a number of these themes in setting out the objectives of telecommunications policy and regulation in Canada. "Telecommunications," the section begins, "performs an essential role in the maintenance of Canada's identity and sovereignty" and, therefore, should be oriented to aims that include the following:

- "the orderly development throughout Canada of a telecommunications system that serves to safeguard, enrich and strengthen the social and economic fabric of Canada and its regions" (s. 7a)
- affordable access to services (s. 7b)
- the national and international competitiveness of Canadian telecommunications firms (s. 7c)
- domestic ownership, control, and use of telecommunication carriage and transmission facilities (ss. 7d, 7e)
- fostering "increased reliance on market forces for the provision of telecommunication services" (s. 7f)
- stimulation of research and development pursuant to technological innovation (s. 7g).

Similarly, the Broadcasting Act (1991) provides for a broadcasting system "effectively owned and controlled by Canadians" (s. 3a) that will, like the telecommunication system, "serve to safeguard, enrich and strengthen the social and economic fabric of Canada" (s. 3.1d[i]). Under this rubric, the act also states that broadcasting, in both its state-owned and private manifestations, should have the following policy objectives:

- to nurture Canadian culture by "providing a wide range of programming that reflects Canadian attitudes, opinions, ideas, values and artistic creativity" (s. 3.1d[ii])
- to bolster domestic creative and cultural industries (3.1f)
- to provide an outlet for "differing views on matters of public concern" (s. 3.1i[iv])
- to adapt to scientific and technological change in a manner that "does not inhibit the development of information technologies and their application" (ss. 5.2c, f).

The statutory objectives of Canadian communication policy are thus broad and encompassing. In some cases, these diverse objectives can be constructed as complementary. For example, the rhetoric of nationalism has been invoked to support unfettered technological development as well as to justify strategies aimed at shoring up the profitability of domestic communication industries. But they can also be seen as conflicting, as the measures required to meet some of these goals can be at odds with those required to meet others. The field of Canadian communication is one of inherent contradictions and enduring tensions: between cultural and industrial policy, between nation building and commerce, between democracy and elite accommodation, between the public good and private interests, and, ultimately, between the state and the market as regulators of the distribution of communicative resources and power. Public policy must navigate this terrain.

The question relevant to this discussion concerns the extent to which the practical policy priorities governing the development of new ICTs in Canada have been substantially democratic, in terms of both process and outcome. The historical legacy of communication policy making in Canada suggests that democratic expectations in this regard are not entirely unreasonable, especially when this experience is compared to other policy areas domestically, or to communication policy elsewhere. In an effort to convince his American audience that the prospect of participatory democratic consultation in communication policy is not an "absurd idea," media critic Robert McChesney (1999, 127) has singled out Canada (specifically, the formalized public debate in the late 1920s and early 1930s that preceded the establishment of state broadcasting) as a counterexample to the "undemocratic historical pattern" of communication policy in the United States. Communications scholar Marc Raboy also points to the Canadian approach to communication policy as something of a model for nations seeking balanced, democratic policy making in the age of globalization: "Among Canada's policy particularities are the principles that communication infrastructures constitute a cornerstone of the national cultural heritage, that the main instrument for carrying out cultural and communication policy is a mixed system of publicly owned and publicly regulated public and private industries, and that *the participation of social groups is a central part of the policy making process*" (Raboy 1997, 191, emphasis added). Emphasizing this democratic particularity, he remarks elsewhere, "As far as the basic legislative and policy framework is concerned, a deep-rooted tradition stemming from the early days of radio in Canada ensures that no major change in the system can be instated, or even seriously contemplated, without public consultation. The transparency and extent of public debates regarding broadcasting policymaking in Canada is unique in the world" (Raboy 1995, 455).

## Communication As a Public Issue in Canada

The tradition of democracy in communication policy in Canada is evident primarily in the history of state bodies established with the specific purpose of engaging public debate and consultation on communication issues. As Stephen McDowell and Cheryl Buchwald (1997, 14) point out, "Any significant reworking of public policies in Canada has, by convention and common practice, been accompanied by extensive consultation processes with many groups in society." These processes have included a variety of royal commissions, task forces, parliamentary committees, and advisory councils. Together, their record testifies to the centrality of communication to Canadian democracy, as well as to the principle of public consultation in communication policy making. A selective sampling of this history provides some benchmarks against which the democratic character of contemporary policy making regarding new ICTs can be assessed.

In 1905 the federal government struck a special committee of the House of Commons, the Select Committee on Telephone Systems chaired by William Mulock, to inquire into the state of telecommunications ownership and service provision under the unregulated monopoly then enjoyed by the Bell Telephone Company. The Mulock Committee sat forty-three times and heard submissions from over fifty witnesses, including private citizens, representatives from alternative service providers, public officials, international consultants, and technical experts. It compiled nearly 2,000 pages of testimony and exhibits, including material that raised the possibility of a state-owned telephone service in Canada. The committee also publicized private contracts between Bell, the railroad companies, and its franchisees, as well as data on the company's profits, ownership, and finances (Babe 1990, 97). According to Dwayne Winseck (1998, 127), these proceedings "set an example for achieving some measure of public influence over national telecoms policy, and suggested that basic questions about electronic communication in Canada were still within the realm of public debate."

Still, the Mulock Committee was not exactly a high point of Canadian democracy. As it became clear that Mulock and his counsel, Francis Dagger, were growing increasingly critical of Bell and likely to recommend a state-owned telephone utility, they were unceremoniously replaced by officials more sympathetic to the company (A.B. Aylesworth, who replaced Dagger as the committee's expert, had been the Bell Company's legal counsel). The result was a nonreport and the shelving of the public ownership option (Babe 1990, 95-9). Nevertheless, the public light the committee's investigations shed on the pathologies of Bell's unregulated private monopoly was instrumental to Parliament's decision in 1906 to commence state regulation of the telephone industry under the auspices of the Board of Railway Commissioners. The effect was to bring a modicum of public interest to bear against the potentially undemocratic outcomes of an unregulated private market (in this case, an abusive monopoly).

The issue of communication was placed before the Canadian public again in 1928 by the Royal Commission on Radio Broadcasting, chaired by Sir John Aird. The issue this time was whether radio broadcasting in Canada should be developed commercially, by private interests, or as a public service owned and operated by the state. The Aird Commission began its work by visiting the United States, Great Britain, and Europe to study competing broadcast system models. It solicited opinions from provincial governments, provoking debate in several legislatures. On this basis, the commission set out three possible scenarios: a system of publicly subsidized private broadcasting networks; a national broadcasting system financed, owned, and operated by the government of Canada; and a system of provincially established and operated radio stations. It then commenced public hearings, which were held in twenty-five Canadian cities spanning all nine provinces (Raboy 1990, 22-9). The commission entertained 164 verbal submissions and received 124 written statements, representing interests including private broadcasters, cultural organizations, francophones inside and outside Quebec, small and large businesses, educators, provincial and municipal officials, professional groups, trade unionists, and a variety of public interest advocates. In its report of September 1929, the Aird Commission declared that "Canadian radio listeners want Canadian broadcasting," and recommended the institution of a state-owned, public service broadcaster for Canada (Royal Commission on Radio Broadcasting 1929, 6).

The Aird Report ignited three years of political activism and public debate (Collins 1990, 54), much of it fuelled by a spirited, ongoing public confrontation between the Canadian Association of Broadcasters, which lobbied against the Aird recommendations and in favour of a private, commercial system, and the Canadian Radio

League, which supported Aird's vision of a national public broadcaster. Led by Graham Spry, the CRL was a coalition of nationalist and progressive intellectuals and groups that extended across lines of class, region, language, and gender. Among its supporters were the All Canadian Congress of Labour, the United Farmers of Canada, the National Council of Women of Canada, and such notable Québécois as Georges Pelletier (editor of *Le Devoir*) and Édouard Montpetit (secretary of the University of Montreal) (Raboy 1990, 31-3).

Along with a concern for the vulnerability of indigenous culture under a commercial system, the CRL's case for state broadcasting was based on a conviction that public communication free of domination by powerful private interests was integral to democracy. Referring to communication as "the heart of democracy," Spry (1931) argued, "There can be no liberty complete, no democracy supreme, if commercial interests dominate the vast, majestic resource of broadcasting." On 26 May 1932, the Canadian Radio Broadcasting Act was passed into law, establishing the Canadian Radio Broadcasting Commission. Along with creating a state broadcaster in Canada, the act established the principle that the spectrum for broadcast communication is a public resource, the exploitation of which ought to be regulated by public authorities in service of the public interest.

The institutionalized public debate surrounding the establishment of state radio broadcasting is not exceptional in the history of communication policy in Canada. Indeed, it is but an early entry in a parade of such exercises. In 1949, motivated by the need to develop a national policy for television, the federal government established the Royal Commission on National Development in the Arts, Letters and Sciences under the direction of Vincent Massey. Described as "the most extensive public discussion of communications in Canada up to that time" (Raboy 1990, 94), the Massey Commission held public hearings across the country and received over 400 submissions from individuals and a wide range of public and private interest groups. The democratic function of mass media was of particular concern to a number of these delegations. The Canadian Federation of Agriculture, for example, argued that a public broadcasting system was "a vital part of the question of whether we are to develop a progressive, responsible and alert democracy. It is vital to our prospects for the development of a citizenship capable of critical analysis and balanced judgment." Similarly, the Canadian Association for Adult Education emphasized the role of a public broadcaster in securing the "democratic principle of free expression," which might be compromised by market incentives under a private system. Echoing this theme, the Federated Women's Institutes of Canada called for "citizens' participation" in regulating broadcast programming, especially with regard to the particular needs of women (all quoted in Raboy 1990, 100). Several delegations also made the case for extending the rights and operations of private commercial broadcasting, including a call to divest the CBC of its authority to regulate its private competitors. The Massey Commission report (issued in 1951) recognized that private broadcasting had a role to play in the Canadian system, but rejected the call for a separate regulator; it affirmed the central and indispensable role of state broadcasting in securing Canadian culture, and recommended expanding its scope to include television.

These are but a few examples of the tradition in Canada of providing institutionalized venues in which citizens and public interest advocates might participate meaningfully in deliberations upon communication policy. A complete list would also include the Royal Commission on Broadcasting (1955), the Special Committee of the Senate on Mass Media (1970), the Royal Commission on Newspapers (1981), and the Task Force on Broadcasting Policy (1986). Indeed, democratic participation in communication policy processes has become a norm in Canada, even in the absence of formal requirements for consultation. In 1995, the Department of Canadian Heritage established a committee to review the mandates of the CBC, the National Film Board, and Telefilm Canada. Although the committee's terms of reference did not require it to conduct public hearings, the Mandate Review Committee held consultative sessions across the country with executives and employees from the institutions under review, independent artists, and business, labour and community leaders. The committee also received 150 written submissions from private citizens, interest groups, and industry stakeholders.

## Communication Regulation and Public Engagement

In addition to public participation in these episodic, large-scale inquiries into various aspects of Canadian communication issues and policy, democratic norms have also been institutionalized on an ongoing basis in the regulatory proceedings of the Canadian Radio-television and Telecommunications Commission. Established in 1968 to assume regulatory authority over public and private broadcasters, the CRTC also, since 1976, regulates telecommunication carriers and service providers that fall under federal jurisdiction. The CRTC is an independent public agency with thirteen full-time and six part-time commissioners who regulate communication industries in Canada pursuant to the objectives set out in the Broadcasting Act and the Telecommunications Act. Its regulatory authority is exercised primarily through granting, revoking, and revising licences to provide commercial telecommunication and broadcast services, as well as the attachment of conditions (such as common-carriage and universal service obligations or Canadian content regulations) to those licences. The CRTC is also responsible for the approval of service rates, standards, and facilities, and oversight of ownership and working arrangements in the telecommunications and broadcasting industries. The statutes that provide the agency with its policy direction and mandate always direct decisions on such matters, but there is interpretive latitude in applying these principles. The CRTC's role is "to maintain a delicate balance – in the public interest – between the cultural, social and economic goals of the legislation on broadcasting and telecommunications" (CRTC 2002a).

Ideally, the CRTC institutionalizes the democratic norm in Canadian communication policy making in a number of ways. In the first place, its mere existence affirms the basic principle that communication is a matter of public interest requiring regulation by an independent public authority, rather than a strictly commercial field in which outcomes can be left entirely to the self-interested, and potentially undemocratic, decisions and actions of private actors in markets. In addition, the CRTC's independence insulates regulatory decision making from the secretive, partisan deliberations of the political executive (i.e., the prime minister and cabinet) and opens it to public scrutiny and participation. As Winseck (1998, 211) puts it, under the auspices of the CRTC, "the cloistered backrooms of telecoms politics [were] to be opened to citizens."

This is the second way in which the CRTC institutionalizes the democratic norm in communication policy making: by engaging the public directly in its proceedings. Observing its mandated obligation to "take into account the wants and needs of Canadian citizens, industries, and various interest groups," the CRTC "invites members of the public to comment on all proceedings related to the delivery of telecommunications services and facilities by the companies it regulates" (CRTC 2002b). This invitation extends to proceedings concerning broadcasting as well. Routine processes of consultation, including public hearings and public notices, allow interested members of the public to make written or verbal submissions pertaining to the commission's consideration of licence applications and renewals, major policy issues, and amendments to its regulations.

Ultimate power over communication policy in Canada rests in the hands of the federal cabinet, by virtue of its effective control over the legislative authority of Parliament. The statutes governing the CRTC also contain provisions allowing cabinet to issue explicit policy directions to the commission, as well as to order the commission to rescind, revise, or review its regulatory decisions. Indeed, in the early 1980s, cabinet overturned several CRTC decisions and undermined efforts to enforce public interest objectives (primarily reduced rates and tariffs) on regulated competition in the telephone industry (Winseck 1998, 198-9). Thus, the specific democratic possibilities of communication policy are effectively limited by the democratic character of the institutions and practices of policy making in Canada more generally.

Nonetheless, the foregoing discussion suggests that communication is somewhat of a special case, in that significant democratic processes have persisted as norms in this policy area. The history of recurring, comprehensive public engagement on major communication issues, and the institutionalization of public consultation in CRTC proceedings, attests to this particularity. To recognize this is not to idealize the role genuine democratic engagement has played in the history of communication policy in Canada, or to suggest that all major decisions in this field have been the product of democratic processes that were adequately inclusive, participatory, and responsive. It is also not to suggest that policy outcomes in this area have been primarily attentive to the public interest or unambiguously beneficial to democratic public life in Canada. Democracy has too often been sacrificed to the perceived industrial, technological, and cultural imperatives of communication policy, and there are too many examples of the public interest distorted, undermined, and ignored, to allow us to be anything other than circumspect in designating communication policy as a distinctively democratic sphere of governance (Winseck 1998, 192-5). But we should not underestimate the significance of the democratic particularity of communication policy making in Canada, especially in comparison with other countries, such as the United States, and with other policy fields within Canada, in which opportunities for sustained public engagement have often been rare. To what degree has the treatment of digital ICTs lived up to the historical legacy of quasi-democratic communication policy making and regulation in Canada?

## Communication Policy and Regulation in the Digital Age

There are many reasons to expect that public policy directing the development of digital ICTs should involve significant, comprehensive, genuine democratic engagement. In the first place, as described above, innovations in communication technology have historically invoked processes of widespread public consultation in Canada. Second, the peculiar technical aspects of these media, and their wide application in multiple aspects of social, political, and economic life, suggests that their impact on the everyday lives of Canadians is likely to be significant. This likelihood recommends engaging as broad a spectrum of the public as possible in attempting to divine the public interest in relation to these technologies. Third, a great deal of the early and continuing development of these technologies has been supported by public funds. Digital infrastructure could therefore be considered a public resource, in which case public accountability and serious, democratic involvement in its direction are not unreasonable expectations. Finally, Canada's historical and practical experience with communication technologies in a continental economic and cultural order strongly suggests that democratic outcomes for this technology will probably require substantial levels of state regulation – which has, in Canada, historically involved institutionalized public engagement – to compensate for the undemocratic outcomes yielded by private interests operating in markets under these conditions. The effects of globalization (which I consider in the next chapter) serve only to escalate and complicate these conditions. They also increase the need to bring democratically directed public authority to bear in determining the development of new ICTs in Canada. For all these reasons, democratic priorities would seem to make a particularly strong claim upon the politics surrounding these technologies.

Again, the question is whether this claim is being heard in the context of the contemporary policy environment. In what follows, I will attempt to show that democracy has not ranked highly among the priorities of public policy on new ICTs. Instead, policy in this area has reflected the priority of unfettered technological innovation and growth, and a complementary determination to develop these technologies in ways that maximize their potential as media of industry, commerce, and economic accumulation. These priorities do not coincide neatly with the democratic public interest. These goals have been pursued vigorously, however, often presented in the language of non-negotiable technological determinism, fatalist concession to the imperatives of global economic competitiveness, and opportunistic, manipulative appeals to Canadian nationalism.

These discourses serve not only to delegitimize alternative possibilities, but also to alleviate the need even to consider them in any serious way, and to obscure the democratic deficit in failing to do so. This has been evident in the manner in which these priorities have been articulated and enforced upon a policy process whose connection to the Canadian tradition of institutionalized democratic engagement in communication policy making and regulation is tenuous at best. Policy making surrounding the new technologies has been characterized by a truncation of opportunities for participation, a historical departure from the inclusiveness of exercises such as the Aird and Massey Commissions, and a consistent tendency to respond more readily and decisively to the interests of major commercial and industrial actors than to those of public interest advocates and their constituencies. In many cases, policy outcomes emerging from these processes have served also to undermine the public interest in enforcing democratic norms on this crucial area of social life. They have done so by routinely surrendering authority to regulate significant matters concerning these technologies to the private calculations and interactions of powerful economic actors in commercial markets, a mode of regulation that cannot be simply equated with democracy.

## Policy in the Digital Coal Mine: Industry Canada and CANARIE

Two events in 1993 were crucial to the early direction of Canadian policy with regard to digital ICTs: the breakup of the federal Department of Communications (DOC) and the establishment of the Canadian Network for the Advancement of Research, Industry and Education (CANARIE). The DOC had been established in 1969 in recognition of the escalating importance of communication to Canadian economic and cultural sovereignty. For the first time, ministerial responsibility for both broadcasting and telecommunications was gathered under the auspices of a single department whose primary concern was "to promote the establishment, development and efficiency of communications systems and facilities for Canada" (Government Organization Act, 1969). It was hoped the DOC could develop a coherent national policy framework for communication that comprehended and assimilated the various tensions alluded to earlier in this chapter. In particular, it was hoped the DOC could harmonize the culturalist objectives that had typically dominated broadcasting regulation with the economic priorities driving telecommunication policy. Looking into "the future prospects of telecommunication in Canada" and responding specifically to the possibility of "a Canadian network of computer/communications systems," the 1971 report of the DOC's Telecommission, entitled *Instant world,* concluded that despite their tradition of separate and sometimes divergent emphases, broadcast and telecommunication policy would have to converge (DOC 1971, vii-viii).

In the ensuing two decades, however, a national telecommunications policy failed to materialize (Raboy 1997, 199). If anything, during this time, the primarily economic goals that had traditionally animated telecommunication policy began to make their way into broadcasting, in what Raboy describes as a "cultural industries approach": "The rationale for public support of content development shifted from nation building to industrial growth ... Cultural nationalism continued to be the predominant theme of policy discourse, but actual policy programs focused on beefing up the industry side of the cultural industries equation" (200). An identification of the public interest with consumer issues (i.e., low service rates, freedom of choice) and of citizens with market actors (i.e., customers and shareholders) also surfaced in this period, as did the rhetoric of technological determinism and industrial competitiveness in global markets. Emblematic in this respect was the view, articulated in a 1981 DOC paper, that "Canada has no choice but to promote vigorously introduction of the new technologies in order to maintain and increase its international competitiveness" (quoted ibid.).

In 1985 the Royal Commission on the Economic Union and Development Prospects for Canada (the Macdonald Commission) issued its report, which – though not explicitly concerned with communication policy – had a dramatic impact upon this sector. The Macdonald Report stressed the need for Canada to maintain an adaptive economy capable of adjusting to international economic changes and new technologies. In line with the neoliberal spirit of governments in the United States and Britain, the report recommended greater reliance on market strategies, diminished state intervention in the economy, free trade with the United States, and rationalization of the Canadian welfare state and social spending. As suggested in the report of the 1986 Task Force on Broadcasting Policy (the Caplan-Sauvageau Report), it was not entirely clear how to reconcile these priorities with a commitment to protect and promote indigenous cultural expression and public service in the broadcasting system, especially since this had historically required greater levels of state intervention and regulation than those contemplated by the Macdonald Commission.

As the 1980s gave way to the 1990s, however, and the diverse potentials of networked computers began to capture the public imagination, the government identified emerging digital technologies of information and communication as a crucial element in Canada's economic future. Harnessing these telecommunication technologies to economic prosperity would, it was believed, require an industrial strategy aimed at supporting the development of digital infrastructure, maximizing the commercial, market potential of new technologies and services, and incentivizing innovation. Thus, some very familiar tensions in Canadian communication policy had once again surfaced: the tension between industrial and cultural goals for communication; between commerce and public service; between market liberalization and state intervention.

In some ways, these tensions reflected an overarching conflict between the priorities of telecommunication policy (historically biased toward industrial objectives) and those of broadcast policy (historically biased toward cultural objectives). Given that the new digital technologies appeared to be erasing some of the distinctions between these two faces of communication, one might expect that crafting a balanced policy for the new media could best be accomplished by a single ministry sensitive, and committed, to both sets of objectives. By 1993, however, the prospect of such a balanced approach was regarded more as a problem to be avoided than a solution to be embraced. The announcement in the United States of a National Information Infrastructure initiative aimed at constructing an "information superhighway" had generated a sense of considerable urgency for a similar effort in Canada. The suspicion was that undue attention on the part of the DOC to the balance between industrial priorities and cultural needs – as well as to the tradition of public consultation in policy making, particularly in broadcasting – might unduly hamper its ability to craft policy directed at developing digital telecommunication infrastructure as rapidly as possible (McDowell and Buchwald 1997, 12).

And so, in 1993, the DOC was dissolved and responsibility for broadcasting and telecommunication was divided: the Department of Canadian Heritage assumed responsibility for national cultural policy and broadcasting; Industry Canada became responsible for telecommunications policy, to shepherd the development of a national digital telecommunication infrastructure. A brief excerpt from Industry Canada's mandate shows how the tensions outlined above were resolved:

> Industry Canada's mandate is to help make Canadians more productive and competitive in the knowledge-based economy, thus improving the standard of living and quality of life in Canada. The Department's policies, programs and services help grow a dynamic and innovative economy that: provides more and better-paying jobs for Canadians; supports stronger business growth through continued improvements in productivity and innovation performance; and gives consumers, businesses and investors confidence that the marketplace is fair, efficient and competitive (Industry Canada 2002).

Pursuant to this mandate, Industry Canada sets out five "strategic objectives": innovation, connectedness, marketplace, investment, and trade. Under the auspices of Industry Canada, policy for the digital age could be shorn of its culturalist and democratic chains, and freed to serve the needs of industrial development and economic accumulation. A crucial question in this respect, as Raboy (1997, 203) has noted, is whether there would "continue to be a public space for the interaction of social, economic and state interests regarding information infrastructure issues."

One of Industry Canada's first acts was to create CANARIE, a non-profit consortium led by major private telecommunications providers. Its initial task was to oversee the upgrade – using a combination of public and private funds – and eventual privatization of CA*net, Canada's original, publicly funded Internet backbone. This was completed in 1997 when ownership and control of CA*net was transferred from the National Research Council to Bell Advanced Communication, for an undisclosed sum. Once this was accomplished, CANARIE's attention turned to two other goals, both of which it continues to pursue today. The first is the construction and operation, again subsidized by millions of dollars in public funds, of a high-speed research and development network connecting universities and the private sector. This project has gone through several iterations, culminating in the $110 million CA*net 4 project, an ultra-high-speed optical network that became operational in 2002. Second, CANARIE funds the development of commercial applications of network technology, concentrating on the fields of remote medicine, distance learning, and electronic commerce.

These developments carry democratic significance. It has been suggested that, with the establishment of CANARIE, "the public interest got left behind" (Gutstein 1999, 82). In the first place, the decision to privatize CA*net was made without public consultation. The public certainly had an interest in this decision: CA*net was a major public telecommunication resource that had been built with public funds, and public funds later continued to flow into network development through CANARIE. Privatization also signalled a significant departure from the model of public access and noncommercial application that had characterized the early development of CA*net and its various regional subnetworks. CANARIE's mission, in contrast, has been decidedly commercial and industrial from the outset. As Donald Gutstein observes, "CANARIE's support for purely commercial enterprises stood in sharp contrast to its lack of interest in the notion of a public network with non-profit uses," and "the threats to public access to information networks that could result from the privatization of the Internet were largely ignored" (94, 88). Although the stakes were high, democratic engagement and public consultation were conspicuous by their absence: "An important, perhaps critical, decision was made about the kind of information networks Canadians would enjoy in the future, yet they had heard nothing about it ... In fact, most Canadians were unaware that major decisions were being made ... These issues were never debated on their merits, at least not in public" (87-93).

The undemocratic character of the new regime directing the development of Canada's digital infrastructure was also reflected in the overwhelming dominance of major industrial or other vested interests within the organization, and the relative exclusion of public interest advocates. Twelve of the fifteen members of CANARIE's first board were from private sector companies with direct interests in telecommunications, including the Stentor group of telephone companies, Northern Telecom, Unitel/AT&T, and IBM, and from the Information Technology Association of Canada, a lobby group representing 1,200 computer and communication firms. Subsequent reorganizations of CANARIE's leadership attempted to strike a better balance between industrial and so-called institutional (i.e., public) members, but as Gutstein (1999, 84) has meticulously documented, "the arithmetic was always questionable." In the first place, major telecommunications, computing, and cable companies have consistently dominated CANARIE's leadership. Second, the interests of members designated "institutional" are not unambiguously public: many are nonprofit consortiums that are themselves dominated by private sector interests, and several represent universities whose links to industry are increasingly more immediate than their identification with the public good. Relatively few genuine public interest advocacy organizations have found themselves represented on CANARIE's board. Thus, "When the public-private balance is recalibrated, the private sector truly dominates," and, instead of a democratic confrontation between public and private interests, CANARIE evinces a more corporatist structure, establishing a "seamless connection between public and private" (85-6).

## The Information Highwaymen

Arrangements and practices such as those described above figure prominently in the constitution of what Gutstein has provocatively characterized as "the private government of connected Canada" (Gutstein 1999, 69). Another institution that has been integral to Canada's policy approach to digital ICTs is the Information Highway Advisory Council. IHAC was established in 1994 by Industry Canada to advise the Canadian government on a comprehensive policy strategy for the development of a national digital communication infrastructure. The government presented IHAC with a list of fifteen public policy issues. The first ones were especially pressing: "How fast should the advanced network infrastructure be built? How will it be financed? ... What is the proper balance between competition and regulation?" (IHAC 1995, viii). The remainder of the list included such issues as Canadian ownership and control, privacy and security, universal access, copyright and intellectual property, and content regulation. In deliberating upon these issues, IHAC was guided by three objectives: fostering job creation through innovation and investment, reinforcing Canadian sovereignty and cultural identity, and ensuring universal access. In addition, IHAC set out five "principles" that would inform its work: an interconnected and interoperable network of networks, public-private sector partnership, competition, privacy and security, and lifelong learning (vii). Thus, IHAC was to be a substantial, institutional consideration of the implications of these new technologies for Canadian society, and of the appropriate policy responses to their arrival. Its magnitude and importance recalled, at least to some degree, bodies such as the Aird and Massey Commissions.

Unfortunately, IHAC departed from the tradition of Canadian communication policy making established by its predecessors in one key respect: its neglect of the democratic imperatives of inclusive participation and public engagement. Over 2,000 people volunteered to sit on IHAC. In the end, the council was comprised of thirty members, nineteen (63 percent) of whom were from the private sector, including eight (27 percent) representatives from major telecommunications, broadcasting, and computing firms (e.g., Bell, IBM, Rogers, Unitel, and Videotron). The remaining members of the commission (37 percent) were from the (nominal) public sector, and included several academics and university or school administrators, consumer advocates, local and regional computer network administrators, a writer, labour leaders and, notably, the chair of the Coalition for Public Information. In its report, IHAC portrayed its membership as representing "a diverse range of interests" (IHAC 1995, ix). This characterization was plausible but debatable, given the near two-thirds majority enjoyed by private sector organizations and the overrepresentation of firms with vested interests in this area, whose ideological convictions were anything but diverse. More serious, in terms of the norm of inclusive and participatory public engagement in communication policy making, is the fact that IHAC held no public hearings and accepted no unsolicited submissions. Instead, IHAC's monthly meetings, and the meetings of its five subsidiary working groups, were held in private. The working groups of IHAC accepted a total of sixty-four invited briefs, but these were not presented in person at a public meeting, and they were made public only in summaries prepared by a consultant hired by IHAC.

Commentators on this process have been nearly uniform in characterizing IHAC as problematic from a democratic perspective. "IHAC's relatively closed style," writes Martin Dowding (2001, 131), "suggests an unwillingness to consult the public." According to McDowell and Buchwald (1997, 4), IHAC's proceedings were "a significant break" from past practices in similar bodies, and could be taken as "a gauge of the receptivity of the government to questions and issues that were not on its agenda, and of the willingness of the government to engage in substantive and significant consultation with a range of groups in Canadian society." Finally, as Gutstein (1999, 115) has observed, "The secretive, closely controlled, corporate-dominated machinations of IHAC contrast starkly with the broad public debate that occurred in Canada in the late 1920s over the commercialization and Americanization of radio airwaves."

This is not to suggest that public interest organizations were not animated by the IHAC proceedings. A diverse array of public interest groups mobilized in response to the IHAC process, including the Public Information Highway Advisory Council, a group named in protest over the lack of public access to the IHAC process (McDowell and Buchwald 1997; Clement, Moll, and Shade 2001). Such groups far exceeded IHAC itself in terms of inclusiveness and participation. For example, the Alliance for a Connected Canada included numerous public interest groups that were concerned about information policy, as well as representatives from a diverse array of labour and citizens' groups. What is also significant from a democratic perspective, however, is just how marginal these groups were to the policy-making process in this instance. Lacking a formal opportunity to participate directly in IHAC deliberations, public interest organizations were forced to organize parallel proceedings aimed at raising public awareness, or to try to influence potentially sympathetic IHAC members through informal, ad hoc channels. The effects of these noble, but marginal, efforts have been well documented: "These activities strengthened the connections between the groups. They have clearly articulated a broadly shared vision ... and had some influence on the wording of official policy recommendations. However, there has so far been *no discernible effect on actual policies or practices,* and there is little prospect that further efforts in the same direction will change the situation" (Clement, Moll, and Shade 2001, 43, emphasis added).

Such a radical departure from the tradition of widespread public consultation in communication policy making could not have occurred by accident. It was never IHAC's intention to consult widely.

The council's director, David Johnston, was adamant that "the mandate of the Council did not call for a broadly-based public consultation process" (quoted in Dowding 2001, 127). When he announced IHAC's formation, Industry Minister John Manley explained, "We want to hear from them in a candid way. We want the flexibility of an *in camera* discussion" (quoted in Austen 1994). In the name of flexibility – which, in this context, could mean only licence to articulate and enforce in private positions feared indefensible in public – major decisions directing the shape of Canada's encounter with digital technologies were made via a process designed explicitly to exclude or minimize public consultation.

Given this, it is not surprising that "from the outset, communication scholars and national public interest group spokespersons worried that the council was dominated by primary stakeholders in the broadcasting, cable and telecom industries, and that issues concerning equity, democratic participation, social justice and employment could be compromised by council member interests" (Walters 2001, 70). These worries were justified. The first final report issued by IHAC recommended exactly what the major industrial stakeholders constituting the bulk of the council's membership had hoped for: an approach that affirmed the undeniable urgency of facilitating technological development, in which primary control over the specifics of this development would be handed over to the private sector.

This outcome was so unresponsive, and potentially damaging, to nonindustrial concerns that IHAC's token labour representatives felt compelled to signal the illegitimacy of the entire process. Rod Hiebert, president of the Telecommunications Workers Union, withdrew from IHAC and asked that his name be withheld from the final report, "because of the council's adamant refusal to address a range of social issues" (McDowell and Buchwald 1997, 22). Jean-Claude Parrot of the Canadian Labour Congress remained on the council, but issued a dissenting minority report arguing that IHAC had failed to address the public interest adequately, particularly with respect to issues surrounding information technology, work, and employment, largely because it was dominated by industry (IHAC 1995, 215-27).

The council made over 300 recommendations. First among them was a concession: when it comes to digital information and communication technology, the market should rule. "In the new information economy," wrote IHAC (1995, ix), "success will be determined by the marketplace, not by government. Hence, the primary role of the government should be to set the ground rules and to act as a model user to inspire Canadians. The private sector should build and operate the Information Highway." The overriding aim of policy and regulation in this area should be "to encourage investment, economic growth, and the creation of jobs for Canadians" (x). Accordingly, "Fair and sustainable competition should be the driving force behind the Information Highway; regulation should ensure an open market, a Canadian presence and a fair game" (xviii). The responsibility of government, in this vision, is to "foster an environment in which the private sector can be innovative and create wealth and jobs" (xix).

These recommendations stand in sharp contrast with the view – arguably more typical of the Canadian tradition in this field – that "policy's role is to hold markets accountable to the obligations which enforce citizens' entitlements" (Abramson and Raboy 1999, 783). Even in relation to such objectives as universal provision of telecommunication services and promotion of Canadian content in broadcasting, where meeting the public interest has historically required relatively high levels of interventionist regulation, IHAC envisioned a shift toward market principles. Thus, with regard to access to services, IHAC's recommendation was, "Public policy should rely primarily on competitive market forces to ensure universal access to Information Highway services at affordable prices" (IHAC 1995, 171). With regard to content, IHAC concluded that "the success of Canadian content will ultimately depend more upon its commercial viability and less upon regulatory obligations" (123).

The council did not completely ignore the need for state intervention to address market failures, nor did it reject outright the regulatory function of the CRTC. IHAC did, however, make it abundantly clear that the development and control of new ICTs and services in Canada was best understood as an industrial project, which could best be achieved by liberating private actors in markets from obligations that might undermine innovation and growth. In this scenario, state authority could be exercised most judiciously by leaving profitable enterprise to the private sector, by creating a regulatory climate conducive to investment and competitiveness, and by spending public money to pick up the slack when market forces failed to meet public interest objectives such as universal access.

These priorities were confirmed in the federal government's initial response to the IHAC report, in which the government pledged itself to "building Canada's Information Highway by creating a competitive, consumer-driven policy and regulatory environment that is in accord with the Canadian public interest and that is conducive to innovation and investment by Canadian industry in new services on the Information Highway" (Industry Canada 1996, 2). The same priorities were reiterated in IHAC's second final report, issued in 1997 with very little additional public consultation. These are exactly the priorities we might expect to emerge from a process designed to be responsive to a very narrow range of powerful private interests: a process that was more exclusive than inclusive, and offered only very limited opportunity for public participation and deliberation, to use terms central to the Democratic Audit. This conclusion is disappointing because IHAC could have been another instalment in the Canadian tradition of broad, democratic, public engagement in communication policy making. Instead, it was a determined departure from this tradition.

Of even greater concern is the fact that IHAC appears to have established a precedent when it comes to making policy for new ICTs in Canada. In 2000 Industry Canada established a National Broadband Task Force (NBTF) to advise the federal government on approaches to making high-speed Internet access available to all Canadians by 2004. David Johnston, formerly the chair of IHAC, was appointed to chair the NBTF. In his preface to the NBTF's 2001 report, Johnston described the task force as "an eclectic group from public and private sector backgrounds ... united by a desire to build a better country by absorbing and focusing ideas and experience from all our citizens"

(NBTF 2001, i). Of the NBTF's thirty-four members, twenty-three were from the private sector, including several representatives from Canada's major telecommunications, computer, cable and Internet service providers: BCE Inc., AT&T, Shaw Communications, Rogers Cable, IBM, and Aliant Inc. There were also eight representatives from the education sector, the president of CANARIE, the coordinator of a nonprofit Aboriginal Internet service provider, and the counsel to the Public Interest Advocacy Centre. Also involved were seven "participating associations," five of which represent industrial interests in the communication sector. The NBTF received sixty public submissions via mail or the Internet, but held no public hearings. Instead, it deliberated privately in five meetings held between January and May 2001. Once again, what might have been an opportunity to engage Canadians in an open, inclusive democratic deliberation about the public interest was reduced to an exclusive, private conversation among vested interests.

The NBTF recommendations, issued in June 2001, reflect precisely the priorities of what has been described as "a techno-economically based policy group ... [whose] dominant membership was senior executives of telecommunications companies" (Dowding 2001, 232). The report defines broadband service specifically in terms of its ability to deliver "full-motion, interactive video games and movies on demand ... [and] improved versions of services available on the Internet today, such as home shopping, electronic banking and electronic newspapers, as well as new residential services such as video telephony" (NBTF 2001, 2). The report also refers to educational and health care applications, but economic uses are clearly primary. On this basis, the NBTF presents the rapid development of broadband infrastructure as a social justice issue of utmost national importance and urgency: "The task force is convinced that using broadband to help bridge the economic and social gaps that currently separate Canadian communities is more than a policy imperative – it is a new, national dream that could bring immense benefits to all Canadians, if we have the courage to live the dream" (3).

In the spirit of CANARIE and IHAC, the NBTF concludes that realizing this dream would demand public subsidy at levels measured in billions of dollars, and also the surrender of effective control over deployment and application to private interests and market forces. The NBTF report recommends that "the private sector should play a leadership role in the development and operation of broadband networks and services," while "governments should facilitate the deployment of broadband networks, services and content through policies and regulations that favour private sector investment, competition and innovation" (NBTF 2001, 4). Governments will have to foot the bill "to deploy broadband infrastructure to communities unlikely to be served by market forces alone" but, in so doing, they must be "guided" by "the value of open, competitive markets," especially once that infrastructure is in place (5).

These recommendations do not coincide neatly with the public interest. They are the predictable outcome of an undemocratic process, dominated by representatives of the very industries that stand to gain most from these policy priorities, in which the public interest was denied an effective, independent, self-determined voice. Indications are that, in the age of the information highway, this dynamic has become the new norm in high-level communication policy making. If this is true, it should represent a source of serious concern for democratic activists and public interest advocates.

## Regulating Digital Communication

In 1994 the federal cabinet – partially to compensate for the lack of public access to the IHAC process – directed the CRTC to gather information and recommend a regulatory framework for managing competition in the communication sector in light of the technological convergence of previously distinct media platforms. Historically, in what is customarily referred to as the separation of content and carriage, firms that controlled access to common carriage media (e.g., telephone companies) had been prevented by regulation from holding broadcast licences or owning broadcasting operations. The reverse also pertained: broadcasting companies could not hold licences to provide common carriage services unless they divested themselves of their broadcast holdings. This principle, borrowed from transportation regulation, was initially applied in a communication context chiefly to meet the economic objective of securing fair competition in the telegraphy market. Eventually the separation of content and carriage developed into a crucial democratic principle of communication regulation, aimed at ensuring that those who controlled access to the wires could not exert undue influence over what passed through them (Winseck 1998, 100-3).

The logic of the content/carriage separation was bolstered by the fact that, for most of the twentieth century, Canadian telecommunications services were provided on a regional monopoly basis, and by the relatively clear technological distinction between point-to-point telephone service and point-to-mass broadcasting. In the 1990s, both these situations changed. The CRTC introduced full competition in the market for local telephone service in 1994 (as had been done for long-distance services in 1992). Meanwhile, the Internet and its related applications seemed to blur the line between telecommunication and broadcasting. Thus, a reconsideration of the content/carriage separation was warranted.

The Order-in-Council (PC 1994-1689) requesting the CRTC proceeding called for "a thorough airing of views" (CRTC 1995, app. 2, i). To this end, the CRTC received 1,085 written submissions and heard 78 oral submissions at public hearings held in Ottawa over three weeks in March 1995. Not surprisingly, most of these submissions came from telephone, cable, broadcast, and entertainment companies, and labour and consumer groups, who typically intervene in CRTC proceedings. But just as significant was the engagement beyond this constituency: "A unique feature of this hearing was the spontaneous flood of submissions from newly-formed public interest groups and activists frustrated until now by the closed-door process established by IHAC. A flurry of grassroots activity resulted in submissions and requests to address the Commission from numerous Canadians who had never before appeared at a hearing" (Clement, Moll, and Shade 2001, 29). Groups such as the Coalition for Public Information, Telecommunities Canada, and the Alliance for a Connected Canada made thoughtful, politicized submissions that stressed the need to moderate industrial, market-based approaches with a strong social and public service model in developing Canada's digital information and communication infrastructure. In sum, the CRTC convergence hearings were far more inclusive and participatory than anything IHAC had attempted.

Was the outcome of the convergence hearings responsive to the inclusive participation for which they provided a venue? The CRTC's report, issued in 1995, affirmed – at least rhetorically – the familiar tension in Canadian communication policy between market- and state-led regulation. Thus, the commission confirmed the government's view that "competition in facilities and services is key to the creation of wealth and ideas in the information economy," but also conceded, "Exclusive reliance on market forces to shape the development of cultural products could well jeopardize the continued availability of Canadian voices and ideas on our communications systems" (CRTC 1995, 5). Nevertheless, on the key issues of cross-platform convergence and the shape of competition, the report was unambiguous. The new regulatory framework would consist of two main elements: "the move to platform reconvergence, and a national champions strategy backing concentration in order to foster homegrown corporations capable of competing globally" (Abramson and Raboy 1999, 781-2). The separation of content and carriage was to be abandoned, allowing telecommunications and cable services providers to own broadcasting operations and vice-versa. Competition *within* the Canadian market was redefined in terms of the global competitiveness of Canadian firms, which necessitated allowing unprecedented levels of concentration, consolidation, and cross-ownership in the Canadian communication sector. This paved the way for a number of subsequent CRTC decisions allowing mergers and consolidations across the telecom/cable/broadcast/Internet/publishing spectrum. Consequently, Canada now has, in terms of ownership, one of the most highly concentrated mass communication sectors in the industrialized world (Hannigan 2001).

This outcome reflected the government's neoliberal industrial policy agenda for the information highway, as well as the interests of the major industrial players involved. This outcome cannot, however, be characterized as responsive to the broad range of social concerns expressed in the considerable process of public engagement that preceded it. Indeed, the CRTC's decision has been described as "confounding several public hearings interveners concerned with opinion pluralism" (Abramson and Raboy 1999, 781-2). More broadly, Clement, Moll, and Shade (2001, 29) write, "Despite the unusually high level of public interest and activity seeking a unique national vision of the evolving information and communication infrastructure, the CRTC did not risk diverging from the official path established by Industry Canada, which, in the end, has the power to overturn any CRTC ruling."

It is not entirely surprising that the CRTC was unable to respond effectively to the breadth of perspectives and interests before it on this issue. Despite its independence, as a regulatory agency the CRTC is basically limited by the policy it is called upon by cabinet to implement and enforce. In this case, cabinet's policy priorities were clearly set out in the Order-in-Council that established the commission's terms of reference for this proceeding. These included specific statements of the government's priority to enable rapid development of "an interconnected and interoperable network of networks," as well as pointed reminders that it was "Government policy" to "foster fair competition and an increased reliance on market forces in the provision of facilities, products and services," and to "update Canadian ownership rules for broadcasting licensees to encourage the investment required to accelerate the implementation of the advanced technologies." The Order also implicitly endorsed the possibility that a single entity could be allowed to act "both as a broadcasting entity ... and as a Canadian carrier" (CRTC 1995, app. 2, i-xi). Accordingly, the CRTC was constrained at the outset by clearly established policy priorities from which it could not deviate, even in response to the alternative priorities expressed at its own hearings.

This raises the question of why inclusive public participation, which had been conspicuous by its absence from the potentially more open-ended IHAC process, was reserved for a CRTC proceeding, the focus of which was more narrow, and whose potential for responsiveness was limited by the fact that most of the major questions had been effectively answered in advance. As Abramson and Raboy (1999, 780) point out, it is strange that a regulatory body such as the CRTC, limited to *applying and implementing* public policy, would hold more extensive public consultations than an advisory body that had ostensibly been created to *advise* the government on the *determination* of policy pursuant to the public interest. Perhaps the staging of extensive public consultation in a process doomed from the outset to be unresponsive was part of a cynical strategy to apply a veneer of democratic legitimacy to a broader policy process that was, on the whole, basically undemocratic. Gutstein (1999, 73) takes this view, describing the CRTC's convergence proceedings as "show hearings." If this is even remotely true, it represents an historic departure from the CRTC's role in providing a meaningful site for effective democratic participation in communication policy and regulation in Canada.

Similar concerns might apply to the CRTC's approach to content issues in light of technological convergence. In its order to explore issues of competition and convergence, the federal cabinet had also requested that the CRTC consider the question of how best to protect and promote Canadian cultural content in the new technological environment. Under the traditional broadcasting policy paradigm, Canadian content is protected through a "delicate balance between obligation and support," in which the state supports domestic cultural production with subsidies of various kinds and, as a condition of their licence to profit from the limited public resource of the broadcast spectrum, broadcasters must include Canadian content prominently in their programming (CRTC 1995, 29). Digital media, however, promise "virtually unlimited capacity, where control is increasingly in the hands of the consumer," and so potentially remove the technological justification for enforcing public service obligations through Canadian content regulations. Interactivity, the asynchronous nature of Internet communication (i.e., unlike a broadcast signal, most Internet content can be accessed at any time), and technological and ownership convergence could also make it difficult to decide whether a given enterprise or service should fall under the statutory provisions of the Broadcasting Act (and so be subject to content regulation), or the Telecommunications Act (and so be free of content requirements).

The CRTC's answer to these questions in its 1995 convergence report was complex. The commission suggested that the medium matters less than the type of activity it mediates: an enterprise's practices should fall under the rubric of the Broadcasting Act in relation to those of its services that clearly constitute broadcasting, while carriage services provided by the same enterprise should be subject to the Telecommunications Act. Nevertheless, the CRTC was also driven by the desire to craft a regulatory framework for digital media that would encourage innovation and investment in new services, and worried that application of the Broadcasting Act's more burdensome obligations to emerging categories of service might serve as a disincentive to their development (CRTC 1995, 31). Since it had already been determined that the market should direct development in this area, the task was to devise a strategy whereby the CRTC's regulatory authority over broadcasting could be *preserved* without being *applied* in a way that undermined the vitality of the new media market.

The commission proposed two possible solutions. The first involved redefining the terms of the Broadcasting Act. The commission pointed out that "the current definitions in the Broadcasting Act will likely capture many new and emerging services that would not contribute materially to the achievement of that Act's objectives" (listed on p. 29 of this text) and that "the development of certain new and emerging services could be expedited by amending the Broadcasting Act to exclude such services from its ambit" (CRTC 1995, 30-1). Specifically, the Broadcasting Act defines a "program" – a designation that invokes content regulation under the act – as "sounds or visual images, or a combination of sounds and visual images, that are intended to inform, enlighten or entertain, but does not include visual images, whether or not combined with sounds, that consist predominantly of alphanumeric text" (s. 2[1]). This definition potentially encompasses a broad range of emerging forms of digital, web-based content. Fearing that the spectre of regulatory obligations might dissuade firms from developing new services and bringing them to market, the CRTC recommended that the definition of a "program" under the Broadcasting Act be amended "so as to exclude, in addition to predominantly alphanumeric text, other services that, while they likely fall within the definition of broadcasting, will not foreseeably contribute materially to the achievement of the Broadcasting Act's objectives" (CRTC 1995, 30). In particular, the CRTC was eager for a definition that excluded "online commercial multimedia services."

It is interesting to note, from a democratic perspective, the commission's admission that "at the hearing, the *majority* view supported an expansive interpretation [of broadcasting and programs] and concluded that there was no compelling need to review the definitions" (CRTC 1995, 30, emphasis added). Thus, the CRTC recognized majority opposition to tampering with the act's definitions, but disregarded it. This suggests that the voices of those private industrial interests that would benefit most directly from such a deregulatory move were loud enough, even in a minority position, to drown out the voices of the majority on this question.

The second possibility explored by the CRTC (upon specific direction from cabinet) was to expand the application of the power it already enjoyed under the Broadcasting Act (s. 9[4]) to exempt broadcasters from particular regulations in cases where compliance was not deemed to contribute substantially to meeting the objectives of the legislation. This would entail outlining specific criteria for a priori exemption that would enable the commission to exempt services in a "more streamlined process [than] the usual public process [used] for more significant applications" (CRTC 1995, 31). In earlier decisions, the CRTC had already issued specific exemptions for video games and home shopping services on the grounds that these services did not contribute to the cultural objectives of the Broadcasting Act; a set of general criteria would allow other emerging categories of service to be exempted automatically, without the need for cumbersome public deliberation on a case-by-case basis. The commission noted that "some parties" expressed concern that such a regime might amount to an end-run around normal licensing procedures for new services, procedures in which opportunity for public comment is required by statute. Substituting a priori regulatory exemption for licensing hearings on specific applications would effectively eliminate a crucial site of democratic engagement in the regulation of new communication services. Nevertheless, the commission recognized that "many interveners also suggested that the exemption process would encourage experimentation and hasten the introduction of new multimedia services" (31).

Reflecting on this divergence of opinion, the CRTC stated its sympathies clearly: "The Commission strongly agrees that the regulatory framework must encourage, rather than hinder, the introduction of new products and services on the information highway" (CRTC 1995, 31). To this end, the commission preferred the first option described above: an amendment to the Broadcasting Act that would redefine its terms so as to "take such services entirely outside the ambit of the Act." This option was not pursued by the federal government, which had hoped that its agenda to relieve digital communication from regulatory burdens and turn its development over to the market could be accomplished within the framework of the existing legislation. Opening up the Broadcasting Act – which would most certainly occasion widespread, lengthy public engagement and force open, democratic consideration of a broad range of possible approaches to technological development – was precisely what the government did *not* want, a fact that had been made abundantly clear in the IHAC process. As such, amending the Broadcasting Act was simply a nonstarter. Alternatively, the commission was open to expanded application of its exemption authority, but felt that further study was required to determine the criteria to be used. For this purpose, the CRTC proposed that another public process be undertaken in order "to determine whether certain categories of services can be dealt with in a more expeditious fashion" (CRTC 1995, 31).

This was the seed and purpose of the proceedings that resulted in the CRTC's *Report on new media,* during which it received well over 1,000 written submissions, and nearly 100 parties made public oral presentations. According to the report, "This proceeding was unprecedented in terms of the broad spectrum of individuals, industries and interest groups from whom the Commission received comments" (CRTC 1999, 3). Once again, the CRTC had provided a forum for democratic engagement in communication policy and regulatory issues that was far more inclusive and participatory than the federal government's IHAC exercise had been. As before, the question is whether the outcome of this process was responsive to that engagement.

The CRTC's primary decision in response to these proceedings was expressed succinctly in its report: "The Commission will not regulate new media activities on the Internet under the Broadcasting Act" (CRTC 1999, 2). In justifying this position, the commission argued, "There is no apparent shortage of Canadian content on the Internet today. Rather, market forces are providing a Canadian Internet presence," and pointed out, "The majority of [hearing] participants [felt] there is no reason ... to impose regulatory measures to stimulate the production and development of Canadian new media content." Both of these assertions are doubtless true. It is less clear, however, that a majority of those who made submissions would therefore also favour a complete forfeiture of the CRTC's regulatory authority in relation to all "new media activities on the Internet." Most public interest groups concerned with communication policy in Canada have consistently expressed strong reservations about an outright handover of the Internet (or any other medium of public communication) to market forces and the powerful private interests that direct them. It is hard to imagine how their submissions in this particular process could have been interpreted as departing significantly from this position. The characterization of majority support for restrained content regulation as indicating generalized support for a broader deregulatory move was at best a sleight of hand and, at worst, duplicitous.

Whether it was responsive or not, the decision to refrain from comprehensively regulating the Internet meant that the development and application of a major new medium of public communication in Canada was left to the private calculations of powerful economic actors in markets, and effectively insulated from democratic, public intervention via the institutionalized regulatory processes of the CRTC. Interestingly, the commission achieved this under the terms of the 1991 Broadcasting Act, without resort to either of the measures – amendment of the act's definitions, or establishment of detailed criteria for a priori regulatory exemption – contemplated in its earlier report on convergence and competition. Instead, the decision not to regulate the Internet was based on three steps that fell squarely within the parameters of the existing act. First, the commission determined that Internet content "consists predominantly of alphanumeric text and is therefore excluded from the definition of 'program' [and] therefore, falls outside the scope of the Broadcasting Act" (CRTC 1999, 8). By this reckoning, most Internet activity was deemed beyond the commission's regulatory jurisdiction.

Second, the commission determined that, in many cases, transmission of information via the Internet – even in forms other than alphanumeric text – does not constitute "broadcasting," as defined in the Broadcasting Act: "any transmission of programs, whether or not encrypted, by radio waves or other means of telecommunication for reception by the public by means of broadcasting receiving apparatus, but does not include any such transmission of programs that is made solely for performance or display in a public place" (s. 2). The commission affirmed that the Internet resembles broadcasting in that it distributes content via "transmission," that personal computers (and related devices) are "receiving apparatus," and that the Internet is not a "public place" in which information is displayed but rather a medium through which it is received. But the commission also pointed out that a great deal of Internet content is "customizable" by consumers who, in effect, create their own unique content by making interactive choices as they gather material on-line. In the commission's view, "this content, created by the end-user, would not be *transmitted for reception by the public.* The Commission therefore considers that content that is 'customizable' to a significant degree does not properly fall within the definition of broadcasting set out in the Broadcasting Act" (CRTC 1999, 10, emphasis added). The argument here is that, for example, websites that include hypertext links to hundreds of connected pages are not transmitting all that information for simultaneous reception by an audience, but rather only providing a means for individuals to customize the content they wish to receive at any given time. Thus, such services do not constitute broadcasting under the terms of the act and so should not be regulated.

What about Internet content that still *does* constitute broadcasting under these assumptions, such as nonalphanumeric content (i.e., sound and images) delivered via the Internet that is not "'customizable' to a significant degree"? The CRTC's third, decisive step was to exempt formally all new media broadcasting undertakings from regulation and licensing requirements. The commission determined that "to impose licensing on new media would not contribute in any way to its development or to the benefits that it has brought to Canadian users, consumers and businesses" and that applying regulation to new media broadcasting undertakings "will not contribute in a material manner to the implementation of the policy objectives set out in section 3(1) of the Broadcasting Act" (CRTC 1999, 11). Without specifying any criteria, the commission moved to issue a blanket exemption order "without terms or conditions in respect of all undertakings that are providing broadcasting services over the Internet, in whole or in part, in Canada." The commission's position is that "new media" simply means "services delivered over the Internet." Therefore, it remains unclear whether the CRTC's regulatory holiday is intended to remain in effect in the event (potentially imminent) that traditional broadcasters begin delivering their programming via the Internet. In any case, on the central question, there is no ambiguity: Canada's major agency responsible for enforcing the public interest in communication through regulation will not be doing so when it comes to the most significant medium of public communication since television. As Françoise Bertrand, chair of the CRTC, put it in announcing the release of the *Report on new media:* "Our message is clear ... the CRTC will not regulate any portion of the Internet" (Bertrand 1999, 2).

## Democracy and Canadian ICT Policy

In 1905, as it became clear that he was likely to recommend nationalization of telephone services in Canada, William Mulock was replaced by Adam Zimmerman as the chair of the government's Select Committee on Telephone Systems. In his opening remarks to the committee, Mr. Zimmerman related the following:

> While on my feet I think it is only right that I should say that the Bell Telephone Company gave us a most delightful outing, which I think was enjoyed, not only by members of the committee, but by members of the House who were with us on Saturday. I know personally I enjoyed it very much. It was not only a source of pleasure, but also a source of very great information, more information I think than we could have received on that particular line here in a month, as far as any witness could give it to us (quoted in Babe 1990, 97).

Prime Minister Jean Chrétien's 2001 golf game with Tiger Woods was apparently not the first time that Bell had organized a "most delightful outing" for those at the highest levels of communication policy making in Canada.

While the Mulock Committee was obviously co-opted, its proceedings still represent the beginning of a Canadian tradition of significant democratic engagement in communication policy making, and of public service regulation of communication industries, that persisted for nearly a hundred years. Communication policy making and regulation in Canada have never been perfectly inclusive, participatory, and responsive, but they have certainly compared well on this score both internationally and relative to other domestic policy sectors. As detailed above, recent observance of the Canadian tradition of quasi-democratic policy making and regulation directed by the public interest has been more rhetorical than material. The emergence of new, digital technologies of information and communication in the 1990s has coincided with a marked departure from the historical norm of institutionalized democracy in communication policy making and regulation. Mr. Zimmerman's picnic and Mr. Chrétien's round of golf are the symbolic bookends of nearly a century of democracy in Canadian communication policy.

The departure from this tradition has been evident at the level of both process and outcomes. In terms of process, the major agencies and advisory bodies driving policy in this area, such as CANARIE, IHAC, and the NBTF, have drastically truncated opportunities for public participation in comparison to similar bodies in the past. Their membership has also been far from inclusive, characterized by a consistent overrepresentation of powerful, private actors with vested interests in this policy area, and only token representation of public interest groups and other constituencies. The public hearings of the CRTC surrounding convergence and new media issues did provide a venue for inclusive participation, but as detailed above, decisions emerging from these processes have consistently appeared as foregone conclusions, suggesting a serious incapacity on the part of the CRTC to respond effectively to the public input it engages. This dynamic of formal engagement coupled with severely limited capacity for responsiveness raises the possibility that the democratic function of the CRTC has been reduced to providing legitimation for democratically suspect policy and regulatory decisions that are taken elsewhere.

The priorities of policy concerning new ICTs have been predominantly economic and industrial, a fact that reflects both the leadership of Industry Canada and the interests of those private actors who have dominated the policy community in this sector. The paramount concern has been to establish a climate conducive to technological innovation, capital investment, and economic growth. As per the orthodoxy of our times, these priorities have charted a course of privatization and regulatory liberalization in the communication sector, particularly in those areas where digital technology plays a role. As we will see in the next chapter, the neoliberal drive in contemporary communication policy has been only somewhat modified by residual concern for the promotion and protection of Canadian culture and, as we will discuss in Chapter 5, for the achievement of universal access to the new technologies and the services they mediate. In most cases, strategies recommended for achieving these public interest goals have complemented rather than moderated the prevailing market orthodoxy: strong industries in competitive markets will produce domestic content and provide reasonable access. In cases where the market fails to yield these outcomes, the state can provide subsidies, but ought to refrain from enforcing regulatory obligations that might act as disincentives to investment and innovation.

This strategy is clearly good for business, but is it good for the exertion of democratic control over the proliferation of these technologies and their applications? Neoliberal strategies of privatization and marketization effectively remove major decisions regarding technological development and deployment from the democratic process and reduce attention to them in public life. The ideology that animates this strategy typically presents the alchemy of competitive markets and choosy consumers as definitively democratic. In this view, freedom from state intervention for corporate actors plus freedom for individuals to choose between packaged options equals democracy. When competition is revealed as a euphemism for concentrated cross-ownership, however, and when consumers act like consumers and make choices based on their private interests rather than as public-spirited citizens, this account becomes very difficult to sustain. More substantially, democracy can be brought to bear on the development of information and communication technology – an increasingly crucial element of social, political, and economic life – through institutionalized opportunities for public engagement in communication policy making and regulation that are genuinely inclusive, participatory, and responsive. Disconnected from its own tradition, recent communication policy in Canada has not only failed to meet this standard, but has also undermined the possibility of doing so any time soon.

## Chapter 2

### Strengths

- Canada has a history of democratic consultation in communication policy making.
- Canada has many groups and actors mobilized around the public interest in communication, culture, and technological development.
- Institutional venues, such as the CRTC hearing process, exist for the engagement of Canadians interested in communication policy and regulatory issues.

### Weaknesses

- Opportunities for public participation in the ICT policy cycle have been systematically diminished.
- A narrow range of industrial interests has been overrepresented in the ICT policy cycle.
- Policy processes that have included opportunities for public participation have been generally nonresponsive.

# 3 COMMUNICATION TECHNOLOGY, GLOBALIZATION, AND NATIONALISM IN CANADA

The message has been scrawled with spray paint on cement walls, and shouted in defiance from burning lungs across steel barricades in the streets of Seattle, Prague, Genoa, and Quebec City: "THIS IS WHAT DEMOCRACY LOOKS LIKE!" It refers to the spectacular demonstrations of conscientious objection that now routinely accompany official gatherings of the vanguard of global capitalism. These demonstrations run the gamut from peaceful street protest to brick throwing; from anarchic raging against the machine to sophisticated articulation of, and mobilization around, viable political alternatives to capitalist globalization. This diversity of activity has a common origin in a democratic impulse: in a democracy, citizens ought to be able to participate meaningfully, as equals, in decisions about the conditions and priorities under which they live together as a society, about how common goods ought to be distributed, and about the content and enforcement of the public interest. The public protests in Seattle, and the public deliberations among civil society activists in Quebec City, demonstrate a widespread and deeply held conviction that core political and economic dynamics entailed in what is loosely called "globalization" fail to meet this standard, and undermine the possibility of recovering adequate grounds for effective citizenship. Capitalist globalization is undemocratic, on this understanding, because it diminishes the power of political institutions (such as national governments) that are directly accountable to citizens, and in relation to which citizens enjoy established rights of participation as formal equals.

Under these conditions, the political control of democratic governments is either constrained by treaty (such as trade and investment agreements) or effectively transferred to powerful actors and institutions (such as transnational corporations and international trade tribunals) that are not accountable to locally defined public interests, and in relation to which few meaningful citizenship rights or opportunities exist. The charge here is that corporate rule, even when codified in agreements freely undertaken by sovereign states and mediated by organizations to which national governments give their consent, constitutes a political condition that is not inclusive, responsive, or participatory enough to be described as democratic. This democratic deficit is aggravated by the fact that decisions to institutionalize this political order have not themselves been subject to adequate democratic consideration in Western liberal democracies, Canada included. Indeed, this direction has not even clearly been democratically *endorsed* (in the 1988 Canadian federal election, the majority voted for parties that opposed a free trade agreement with the United States, but the Conservative government of Brian Mulroney proceeded with the deal anyway). In this light, it might be more accurate to say of recent massive public demonstrations against the expansion of capitalist globalization that this is what it looks like when democracy is systematically denied.

The aim of this chapter is to investigate the relationship between ICTs and globalization, and to explore its implications for Canadian democracy. Issues of technology and communication have always been central to discussion of "the national question" in Canada, and the contemporary dynamics of globalization and digital networks once again bring this discussion to the fore. Does the combination of the political economy of globalization and the rapid proliferation of digital communication technologies create conditions for democratic politics at the national level in Canada that are adequately responsive, inclusive, and participatory? Or does this complex render that possibility more remote?

## Technology, Communication, and Nation in Canada

Technology and communication have been central instruments in the historical imagination of Canada (Anderson 1991). "Technological nationalism," writes Arthur Kroker (1984, 10), "has always been the essence of the Canadian state and, most certainly, the locus of Canadian identity." Maurice Charland (1986, 201) goes even deeper, locating technology "at the center of the Canadian imagination, for it provides the condition of possibility for a Canadian mind."

Technologies such as the Canadian Pacific Railway and, subsequently, the telegraph, telephone, and broadcasting, have been used to craft an artificial nation where before there was none. These technologies operate both materially and ideologically. They provide material links between far-flung compatriots, as well as the language of a synthetic common purpose that can inspire and bind them politically, and perhaps even spiritually. It is not just that "Canada owes its existence to technologies which bind space," but also that "the *idea* of Canada depends upon a rhetoric about technology" (Charland 1986, 199). Technology extends the Canadian presence across territory, and also enforces dominion over the consciousness of its inhabitants. As Charland puts it: "The popular mind, like the land, must be occupied" (206). Because of this imperative, Charland suggests, state-owned broadcasting in Canada was established "to occupy and defend Canada's ether *and* consciousness" (208, emphasis added). Beginning with the Canadian Pacific Railway, and continuing right through to the "New National Dream" of a broadband digital communication network, the material and ideological bases of technological nationalism in Canada have been remarkably consistent. In each instance, technology purports to provide three things:

- a medium of regular intercourse among a population divided by formidable geographic expanse
- a unifying common project that lends coherent purpose to a diverse people, demands their commitment, and provides an avenue through which they might identify with one another
- a means of materializing the sovereignty and independence of the Canadian state and its people.

Communication technologies have been reserved a special place in the mythology of technological nationalism in Canada. In 1970, Harry Boyle, then vice-chair of the CRTC, declared that Canada "exists by reason of communication." Three years later, Communications Minister Gérard Pelletier added: "The existence of Canada as a political and social entity has always been heavily dependent upon effective systems of east/west communication" (both quoted in Babe 1990, 5). Telecommunications technologies such as the telegraph and telephone facilitated the coordination of a national industrial and commercial economy that, in offsetting the strong southward pull of continentalism, was perceived as indispensable to maintaining a modicum of Canadian economic sovereignty. Broadcast technologies are typically presented as having played a similar role with regard to cultural and political integration, mediating a common Canadian experience and psyche, and constituting a crucial site for the exercise and protection of the nation's cultural sovereignty and distinction. Introducing the Canadian Radio Broadcasting Act in the House of Commons in 1932, Prime Minister R.B. Bennett described broadcasting as "a great agency for the communication of matters of national concern and for the diffusion of national thought and ideals ... the agency by which national consciousness may be fostered and sustained and national unity still further strengthened ... a dependable link in a chain of empire communications by which we may be more closely united with one another" (House of Commons 1932a, 3035). More recently, in a passage that captures perfectly the persistent national imperative in Canadian communication, David Taras (2001, 3-4) has written: "Sharing a border with the largest economic, military, and entertainment power on the planet, plagued by deep linguistic and regional differences, and having undergone a series of painful national unity and constitutional crises in its most recent history, Canada must depend on its media system to be a cultural and information lifeline in a way that other countries need not."

In the history of Canadian communication policy, nearly all major documents either begin with or feature prominently a panegyric to the national purpose – whether industrial, cultural, or political – of communication in this country. This perceived national purpose has also provided the impetus and legitimation for many of the policies, programs, and institutions that distinguish the Canadian communication environment:

- overarching national objectives set out in the statutes governing broadcasting and telecommunications
- domestic ownership requirements in these industries
- the maintenance of substantial, state-owned and -operated radio and television broadcasting services
- Canadian content regulations
- Telesat Canada
- the National Film Board, Telefilm Canada, the Canada Council, and a host of other programs and regulations supporting Canada's cultural industries.

With the completion of the CPR, it would seem, communication became, and remained, the privileged site of specifically and self-consciously "national policy" in Canada.

This technological-nationalist orientation of the Canadian approach to communication, however, has not been adequate to the tasks set for it in either its ideological or material aspects. In the first place, though they are presented as tools of independence and national autonomy, communication technologies in Canada have in fact served historically as instruments of cultural dependency and continental integration. Measures such as those listed above certainly represent significant attempts to direct communication technology to the purpose of national consolidation in Canada. But these initiatives have not always resulted in the security of Canadian cultural and industrial sovereignty. "Far more common," writes Robert Babe (1990, 7), "has been the tendency for communication media to be deployed in manners supportive of continental integration; indeed, it may be fairly said that Canada as a nation persists *despite,* not because of, communication media." Babe goes on to document how telecommunication technologies ranging from the telegraph to satellites have been enlisted in the assimilation and coordination of a continental North American economy.

This tendency is even more pronounced in relation to technologies of cultural communication, which have mediated high levels of Canadian consumption of American cultural products. Commenting on this dynamic over two decades ago, critical communications scholar Dallas Smythe (1981, ix) argued that Canada had effectively become a dependency of the United States, and that the primary functions of mass communication media in this complex were to manufacture a Canadian audience, saleable as a commodity to American commercial advertisers, and to produce "the necessary consciousness and ideology to seem to legitimate that dependency." At the dawn of the twenty-first century, the vast majority of cultural products consumed by Canadians continue to originate from foreign, usually American, sources (Taras 2001, 188; Straw 1996, 98; Magder 1996, 150). Mass communication technologies in Canada have served more readily to deliver American culture into Canadian homes (and Canadian attention to American corporations) than they have to provide a medium for expressing and encountering a culture that is indigenous and distinctive. Mass communication technologies have been instrumental in generating a condition in which American culture is no longer experienced by most Canadians as especially foreign. As John Meisel (1986, 152), former chair of the CRTC, put it: "Inside every Canadian, whether she or he knows it or not, there is, in fact, an American. The magnitude and effect of this American presence in us varies considerably from person to person, but it is ubiquitous and inescapable."

The official approach to Canadian content on the Internet has been interesting in this regard. In its first final report, the Information Highway Advisory Council (IHAC) stressed its endorsement of "the Canadian cultural imperative [that] Canadians must be able to provide their own content on the Information Highway" and suggested that "government should continue to have the tools and mechanisms necessary to promote Canadian content" (IHAC 1995, 28-9). In its second final report, IHAC went further, characterizing the situation surrounding "Canadian sovereignty and cultural identity" on the Internet as "urgent" and calling for "expanding the scope of Canadian cultural policy to encompass new media" (IHAC 1997, 59-60).

Despite this position, however, IHAC and subsequent policy measures made it clear that the Internet should not be subject to the content restrictions and obligations typically applied to Canadian broadcasting. In 1995 IHAC pointed out that the Internet was "principally a private communication medium" and that therefore "regulating its content is no more appropriate than regulating the content for the telephone" (28). As suggested above, by 1997 IHAC had come to recognize the need to articulate a cultural policy for the Internet. Still, regulated Canadian content requirements enforced upon enterprises profiting from distributing material via the Internet were never part of the plan. Instead, in its two reports, IHAC laid out a strategy that was essentially industrial in its aims, reflecting the economic logic that animated the rest of the council's work.

For IHAC, the key issue was "the size and financial condition of the companies producing and distributing Canadian content in English and French and the degree to which different types of Canadian content are able to recover their costs from domestic and export sales" (IHAC 1997, 64). Accordingly, "Government initiatives should focus on strengthening the creation, production, distribution and marketing of content made by Canadian creators or addressing Canadian subject matter." Such initiatives would include: government funding and business development assistance for the production of "new services and products for the Information Highway" (68); state support of collaborative ventures between cultural producers, industry, and research communities; a copyright and intellectual property regime that encourages innovation; state-sponsored digitization of public and cultural information (in English and French); and, finally, the maintenance of a relaxed regulatory climate in which Canadian firms might grow to compete in international markets.

This program articulated neatly with the traditional language of Canadian technological and cultural nationalism. It also envisioned considerable state intervention, but none that would involve placing significant public interest obligations on firms cashing in on the "information revolution." As discussed in Chapter 2, this approach to securing Canadian content mediated by ICTs was confirmed by the CRTC in its convergence and new media proceedings. In both cases, arguments about the Internet's technological characteristics (i.e., interactivity, decentralization, abundant bandwidth) were invoked to rule out the desirability and possibility of content regulation. The CRTC's *Report on new media* was particularly categorical in this respect, to the point of rejecting IHAC's assertion that Canadian content on the Internet was a problem requiring state attention. Citing statistics that Canadian sites represented 5 percent of all websites, the CRTC concluded that "the circumstances that led to the need for regulation of Canadian content in traditional broadcasting do not currently exist in the Internet environment. Market forces are providing a Canadian presence on the Internet that is also supported by a strong demand for Canadian new media content" (CRTC 1999, 14). For greater emphasis, the CRTC stressed, "There was no convincing evidence submitted throughout the hearing process to suggest the visibility of Canadian content on the Internet is a problem and, therefore, no policy rationale to pursue regulatory measures to support access to Canadian content on the Internet" (15). Ultimately, this view complemented rather than contradicted the IHAC approach: IHAC had recommended a cultural policy for the Internet that amounted to state subsidy for domestic information industries; the CRTC findings ensured that such a policy would never extend to enforcing public service obligations upon the private firms receiving public subsidy.

That this could all be accomplished in the name of national purpose points to another questionable aspect of technological and state-led cultural nationalism relative to communication: its tendency to undermine the public interest in communication. Technological nationalism is easily co-opted to industrial or commercial strategies that primarily benefit private interests and, moreover, obscures the authenticity and diversity of the Canadian national experience. On this account, the Canadian nation conjured by technology is deeply artificial and tenuous, grounded in nothing more substantial than faith in the virtue of mediated communication itself. As Maurice Charland (1986, 213) puts it, "Technological nationalism offers Canadians a common experience of signs and information in which culture is disembodied ... a cultural experience which is not grounded in a region or tradition." Canadians encounter technology intimately, but what they experience through it is quite remote: common (but passive) consumption of "cultural" commodities, and mass (but solitary) audience to mediated spectacles. A culture so lacking in gravity is easy to appropriate. Empty of substance, technological nationalism acts as a vessel that is readily filled, from above, with material suited to the designs of the industrial and state actors who control technological resources, but which does not necessarily reflect the diverse realities of either Canadian culture or the public interest.

Examples abound in which the potent rhetoric of technological nationalism has been conscripted to justify activities primarily intended to make the Canadian communication market safe for capitalist economic accumulation, and for the fortunes of powerful private interests. As we have seen, this tendency has also characterized Canada's approach to the development of digital ICTs, often under the cloak of national purpose. This technological nationalism is complicated further when it is twinned with cultural nationalism. The public interest has often been invoked to support Canadian communication policy, but its identification with an artificial, state-cultivated national culture has typically served to obscure, and in some cases to repress, the plurality of cultural experiences and interests that have always been present in the Canadian fabric. For example, Aboriginal, Québécois, regional, working-class, immigrant, and women's cultures have not always been well served, or even adequately represented, in the imagination of Upper Canadian cultural nationalists. Furthermore, both in telecommunications and broadcasting, national consolidation has consistently been achieved at the expense of reasonable provincial autonomy in these areas.

To emphasize one among these examples, the misappropriation of Québécois culture has been a consistent, and compelling, criticism of cultural nationalism in Canadian communication. According to Richard Collins (1990, 56): "Assertion of Canadian-ness in public broadcasting against the 'external contradiction' of the United States meant that the public broadcasting system tended to suppress regional differences, provincial interests, and most important the different cultural and historical experience of francophone Canadians." Commenting on the report of the Massey Commission in 1951 – arguably the high-water mark of English Canadian cultural nationalism – André Laurendeau wrote: "What dike can be built to defend against this irresistible tidal wave [of American mass culture]? Only that of a Canadian national culture; that is to say, a myth, phantom, the shadow of a shadow. There is not one Canadian national culture. There are two: one English and one French" (quoted in Raboy 1990, 110). As Marc Raboy points out, "Similar arguments against centralization could have been formulated from the point of view of other regional, cultural or social communities." The number of national (or non-national) cultures against which the conception of a singular Canadian national culture might do violence is well in excess of two. Predictably, Canada's official linguistic duality is well represented in the suite of funding programs developed by the Department of Canadian Heritage under its Canadian Digital Cultural Content Initiative (for example, 50 percent of all funds distributed under the Canadian Culture Online funds are earmarked for French-language material). Despite rhetorical acknowledgment of Canada's "rich diversity," however, the needs of identity groups not captured by the French/English distinction are not explicitly addressed, a significant exclusion in our pluralistic society.

Still, despite the inadequacies of technological and state-sponsored cultural nationalism in Canada, a nationalism of sorts was crucial to the establishment and maintenance of the quasi-democratic regime in Canadian communication policy described in Chapter 2. To recall, for most of the twentieth century, this regime was characterized by a conviction that Canada's cultural distinction and political sovereignty necessitated domestic control over communication infrastructure, by a commitment to public ownership and regulation in communication markets, and by an affirmation of the virtue of inclusive public participation in communication policy making and regulation. The empty vessel of Canadian nationalism has proven as useful a container for these aspirations as it has been for other, perhaps less appealing, political, economic, technological, and cultural designs.

At least one strain of communication nationalism in Canada – a strain, not coincidentally, anchored in civil society rather than industry or the state – has defined the national purpose in communication in explicitly democratic terms. Here, the positive goal is clear: securing the ability of Canadian citizens to participate to some meaningful degree in determining and controlling their communication environment as mediated by mass technologies. We might recall from Chapter 2, in this context, Graham Spry's invocation of self-determined communication as "the heart of democracy." From the outset, this democratic (as opposed to technological or cultural) strain of nationalism has operated under a consistent set of three assumptions. First, self-determined communication requires insulating the Canadian media sphere from the encroachment of American control and influence. Second, democracy demands that the requirements of communication as a public service trump its exploitation as a capitalist industry. Third, achieving the common democratic interest in communication requires state intervention (i.e., ownership and regulation) in the markets that distribute communication resources. For better or for worse, the democratic imperatives of communication are inextricably linked with the politics of the national state and capitalism in Canada.

As Spry famously quipped in a 1932 speech before the Special Committee on Radio Broadcasting, "It is a choice between commercial interests and the people's interests. It is a choice between the state and the United States" (House of Commons 1932b, 45-6). Thus, the issue of capitalist globalization as it pertains to communication has seemingly been on the table in Canada for at least seventy years. As the foregoing discussion makes evident, however, the role of the state in this dynamic has been more ambiguous than implied by Spry's stark dichotomies. For example, state-sponsored technological and cultural nationalism in Canada has not consistently served "the people's" rather than "commercial" interests. It is probably true that, absent considerable state intervention in communication markets, the Canadian mediascape would have been dominated (both technologically and culturally) by American capitalists to an even greater extent than it was throughout the twentieth century. We can also concede that purely market-based mechanisms for distributing communication resources are generally less democratic than state-led political processes. Nevertheless, the spectre of American domination has consistently provided the Canadian state with reflexive justification for communication policies that either serve the interests of domestic capitalists, or bolster the apparatus of an artificial national culture that disguises the cultural complexity of the country's inhabitants.

The effect of the Canadian state's intervention in communication markets has not been unambiguously democratic. Instead, the position of the Canadian state vis-à-vis communication is essentially contradictory. On the one hand, the state has been, and remains, an indispensable bulwark against the undemocratic operation of the capitalist market and the total assimilation of Canadian into American mass culture. On the other hand, the state has been, and remains, a powerful agent committed to the profitability and competitiveness of Canadian commercial industries and to the enforcement of an official, national cultural discourse, both of which have significant democratic liabilities.

The state is thus both a friend and an enemy of democratic communication in Canada. This contradiction must be kept in mind as we confront questions about the prospects of inclusive, participatory, and responsive democracy in the age of capitalist globalization and digital networks. Does the political economy of globalization, in which ICTs are deeply embedded, offer greater or fewer opportunities for Canadians to participate meaningfully, as equals, in decisions about the shape of their communication environment? Is the communication environment structured by these conditions more, or less, inclusive of the plurality of experiences and practices that constitute contemporary Canadian society? And finally, are Canadian political institutions more, or less, capable of responsiveness to indigenous aspirations and priorities under the auspices of globalization?

## Globalization and Democracy

Globalization – the lightning rod of contemporary politics and intellectual discourse – is a phenomenon that admits of many definitions and aspects. Generally, globalization refers to the historical process whereby significant economic, political, and cultural activities are decreasingly contained within the borders of territorially defined states and, instead, flow across these borders with relative ease. The structural configuration of core elements of economic, political, and social life in both affluent and developing countries is now best described, it is argued, as transnational or global rather than national. In economic terms, globalization denotes the process through which "national economies are now integrated into a single global marketplace through trade, finance, production, and a dense web of international treaties and institutions" (Cameron and Stein 2002b, 1). Ownership could be added to this list: the era of globalization has coincided with unprecedented levels of concentration and consolidation in the ownership of capital and material resources by massive transnational corporations, a dynamic particularly pronounced in the media and cultural sectors (Herman and McChesney 1997).

One week before the World Trade Organization's 2003 meeting in Montreal, Pierre Pettigrew, Canada's minister of international trade, averred that globalization is "something that has to be managed, but can't be stopped. It is something that is created by technological developments. It's not something that governments decided they wanted to do" (quoted in Patriquin 2003, 8). Notwithstanding the minister's technological determinism, the transnational reorganization of the global capitalist economy is political in both its origins and its implications. Although often represented as a disembodied, irresistible, and ahistorical force to which the world has had no choice but to adapt, globalization is the result of decisions made by governments in the industrial West. These decisions have aimed to liberalize or relax nationally specific controls and regulations on the flow of capital, and on the production, trade, and marketing of goods and services within and across jurisdictions.

This current of transnational market liberalization has been codified in a string of trade and investment agreements among sovereign countries, such as the North American Free Trade Agreement, the General Agreement on Tariffs and Trade, and the General Agreement on Trade in Services. These arrangements are crafted and enforced by a range of international institutions – including the World Trade Organization (WTO), the Organization for Economic Cooperation and Development (OECD), the World Bank, the International Monetary Fund, and a number of regional political and economic organizations – charged with managing the transnational capitalist economy. Taken together, these agreements and institutions comprise what can be characterized as a "supraconstitution" that transcends the independent national constitutions of contemporary states (Clarkson 2002; McBride 2003). Like any constitution, this one effectively defines the space of politics by establishing the extent of the state's political authority, by institutionalizing power relations, and by determining the possibilities of citizenship.

Under the constitution of capitalist globalization, individual states experience a serious challenge to their ability to manage independently their economic activities according to exclusively domestic priorities. States still make and enforce decisions, but their possible courses of action are increasingly hedged by conditions set by international economic institutions and agreements such as those listed above. This is particularly true in cases where states might consider market interventions that involve placing regulatory constraints or redistributive demands upon private enterprises, or that advantage or protect domestic firms at the expense of foreign competitors. In volunteering for neoliberal globalization, countries such as Canada opt into a political economy in which their autonomy is compromised in several ways: by trade and investment agreements that prohibit certain forms of market intervention, and by the international authorities that enforce these; by transnational corporations whose power, mobility, and leverage have been augmented by liberalization of the global economic environment; and by the perceived imperative to provide as hospitable a climate as possible for capital investment and accumulation. Increasingly, moreover, the legitimacy of governments exerting authority over, and on behalf of, citizens within fixed territorial boundaries is challenged by the rise of nonterritorial identities and cultures mediated by digital technology, as well as by the increasing importance of cities as nodes in global economic and informational networks (Gibbins 2000, 672-3).

Nevertheless, the eclipse of the sovereign state by globalization should not be exaggerated. As David Cameron and Janice Gross Stein (2002b, 1, 8) point out, "Obituaries for the state are premature ... Global markets and global politics are certainly expanding, but they do not constrain the state from fulfilling its social contract with its citizens. States still have real and significant capacity, both to provide public goods to their citizens and to mediate the impact of global economic, social and cultural forces." In this respect, the central question may be whether states choose to exercise this capacity, and to what extent the material and ideological conditions of capitalist globalization encourage or discourage this choice. That being said, many people view neoliberal, capitalist globalization as the economic base of a political condition that is profoundly undemocratic.

Globalization can be understood as undermining democracy in two respects. First, neoliberal globalization enforces the protection of private economic activity in markets from constraint by localized, democratically accountable political authorities acting in the public interest. In this sense, globalization is undemocratic because it is *depoliticizing:* it removes decisions about the distribution and regulation of social resources and practices from the public, political realm, and surrenders these to the calculations of powerful, private actors in markets. Admittedly, liberal democratic states such as Canada have never been perfectly committed to bringing the public good to bear against the private interests of capitalists, and have often acted as instruments for securing the conditions of their continued power. Still, in capitalist economies, hope for mitigating the unbridled rule of economic elites in their own private interest generally resides in the institutionalization of formally democratic, and more or less effective, structures of participation, representation, and accountability in national governments.

This suggests the second democratic shortcoming of capitalist globalization: the absence of satisfactory, formalized mechanisms of democratic participation, representation, and accountability in the powerful institutions – international agencies and transnational corporations – that increasingly determine many of the conditions under which people live. No one is a citizen of the WTO or Sony; the OECD's constituency is the political and economic elite of the world's most powerful countries; and Nortel is accountable only to its shareholders. Under this regime, the locus of effective power is seriously disconnected from the practices of citizenship, representation, and accountability. It is in this disconnection that globalization emerges as a potential democratic crisis.

## Globalization and Communication

The relationship between globalization and information and communication technologies is intimate. ICTs are instrumental to the operation of the global economy, providing an infrastructure for the execution and coordination of economic activities (i.e., production, consumption, trade, finance) that are territorially disaggregated and dynamic (Castells 2001, 64-115). The centrality of ICTs to global capitalism is especially pronounced with regard to the service and knowledge-based industries that play an increasingly dominant role in this economy (Deibert 1997, 137-76). New media technologies are also indispensable to the production and circulation of the branded entertainment and information commodities contributing to the maintenance of a global commercial culture that transcends national distinctions (Hannigan 2002; Klein 2000). As will be discussed in Chapter 4, technologies such as the Internet have also been crucial in mediating the activities of transnational social movements, and provide a communication infrastructure for the nascent global civil society that has emerged in parallel with the globalization of the world economy (Deibert 2002a). In many respects, then, globalization and ICTs are inextricably linked.

Perhaps most significant from the perspective of this audit, however, is the manner in which globalization affects the prospect of exerting democratic control over the development and configuration of these technologies, and of directing them to democratic purposes. New ICTs are not only instrumental to the dynamics of globalization described above; they are also subject to these dynamics. Globalization establishes the conditions under which public policy can be formulated and applied to digital technologies and the practices they mediate. Whether communication policy making and outcomes in Canada can meet robust standards of democracy depends on whether the Canadian state is willing and able to carve out a self-determined policy approach to these technologies that is responsive to the priorities and needs of Canadian citizens. At a minimum, such an approach would also have to be capable of moderating the market forces and private interests that have historically been inadequate to the task of ensuring a democratic distribution of communication resources in this country. As discussed in Chapter 2, recent communication policy in Canada has generally failed to meet this democratic minimum. To what extent is this failure linked to, or even caused by, globalization?

Mass communication mediated by technology has always required some level of international cooperation and agreement (Karim 2002, 273-9; Raboy 2002, 125-7). In 1865 the International Telegraph Union was established to coordinate the transmission of electric telegraph signals between countries; ten years later, the Treaty of Berne created the General Postal Union. In 1886 an international convention on copyright was signed, the forerunner of the World Intellectual Property Organization, established in 1974. Conferences in Washington in 1927, and Berlin in 1932, established a regime for distributing radio frequencies among the regions and countries of the world and created the International Telecommunication Union to manage this regime. In 1964 the International Satellite Organization was established to manage the international administration of satellite telecommunication.

A significant feature of the original regime in global communication regulation was the extent to which it achieved international coordination without sacrificing the sovereign authority of states to determine independent policy approaches within their respective jurisdictions (Raboy 2002, 125). International conventions were required to manage communication resources and practices that either straddled or crossed national borders (e.g., the broadcast spectrum or telegraphic messages), but these were merely an adjunct to national sovereignty over the use of those resources (126). This regime thus enabled, for example, the proliferation of distinctive models – some private, some public, and some hybrids – for organizing and regulating broadcast communication in various national settings.

The basic assumptions of this regime persisted until the wave of privatization, regulatory liberalization, and transnational consolidation that swept communication sectors throughout the industrialized world in the 1980s and 1990s, although its erosion had begun decades earlier. The United States, bolstered by its elevated political and economic status in the wake of the Second World War, began to promote vigorously the idea of communication and culture as commercial commodities and the complementary doctrine of a "free flow" of information across national boundaries. The latter was essentially an attempt to secure conditions whereby American media enterprises and products could penetrate foreign markets free from prejudicial regulatory resistance.

By the 1970s, developing countries were beginning to suggest that the free flow principle might just be a liberal democratic veneer over a less palatable strategy of cultural imperialism being carried out by powerful actors in the industrial West and North. These countries began to press for a "New World Information and Communication Order" (NWICO), in which the ability to self-determine domestic communication and cultural environments would be recognized as fundamental to development and independence. In 1978 the United Nations Educational, Scientific and Cultural Organization issued a Mass Media Declaration that gave voice to some of these concerns. In 1980 UNESCO's International Commission for the Study of Communication Problems, chaired by Seán MacBride, issued a report that "assailed the one-way flows of information from North to South and the constraints on communication imposed by commercialism, advertisers and media concentration" (Karim 2002, 279; UNESCO 1980). Soon after, UNESCO formally resolved to work toward a NWICO and a better deal for less powerful countries in the global communication environment – a move that prompted the United States and the United Kingdom to withdraw from UNESCO.

This withdrawal was a harbinger of the new international regime under which policy and regulation regarding emerging ICTs such as the Internet would be elaborated. The features of this regime correspond neatly with the exigencies of neoliberal globalization:

- privatization of communication industries and resources
- liberalization of communication regulation and a corresponding move to market-based distribution of content and services (i.e., competition)
- consolidation and concentration of ownership in communication and technology sectors, both domestically and transnationally
- relaxation of constraints on international trade and investment in communication goods and services.

Justification for this regime shift typically combines dogmatic acceptance of the free market as the only legitimate mechanism for distributing social goods with determinist surrender to the imperatives of converging technologies and a "borderless" world (Deibert 2002b). Its goal is customarily set out as providing a climate for economic growth (including, especially, job creation) through technological innovation and investment, an aim that is routinely presented as a condition of national autonomy. This characterization is especially cunning, given the diminished will and capacity of sovereign states (and, by extension, of the citizens that authorize state power) to determine their own technological and communication environment under this regime. Much of this capacity has effectively been transferred to massive private corporations that develop and distribute social resources such as communication and technology in open markets according to calculations of self-interest, and to the vanguard international institutions that keep the global capitalist economy safe from the meddling of interventionist states. Indeed, the primary end of this regime is arguably to open public service-oriented national communication systems to exploitation by private, transnational capital.

## The Globalization of Communication and Canadian Democracy

To secure the global regime described above, the role of the state in the market had to reconfigured. Additionally, the understanding of communication as a democratic, public good closely bound up with political and cultural autonomy had to be replaced by a conception of communication as a profitable industry with "products" – information, culture, services – that were private property and marketable commodities (Schiller and Mosco 2001, 13-14). As discussed in Chapter 2, by the 1980s, Canadian communication and cultural policy was clearly moving in an industrial, market-oriented direction.

This direction cannot be understood apart from Canada's voluntary assimilation into the political economy of globalization. In the closing decades of the twentieth century, Canada became a party to a series of agreements: the 1989 Canada-US Free Trade Agreement (FTA); the 1993 North American Free Trade Agreement (NAFTA) among Canada, the United States, and Mexico; and the 1993-4 General Agreement on Trade in Services (GATS) and 1997 Basic Agreement on Telecommunications, both signed by members of the World Trade Organization. As Dwayne Winseck (1998, 216) writes, "Although regulatory liberalization predates free trade, there are basic links between changes in Canadian communication law and the FTA, NAFTA, WTO trilogy. These agreements harmonize communication and investment policies by diminishing constraints on transborder data flows, restricting the scope of public sector activity, and limiting the range of telecoms services that can be publicly regulated." Despite the significant differences among these three agreements, several basic elements relevant to matters of communication are consistent throughout, and exemplify the challenge to democracy posed by globalization in this respect.

The basic operation of these agreements is fairly simple: signatories pledge not to engage in market interventions (such as tariffs, directed subsidies, state ownership, tax concessions, and discriminatory regulations) where such action will result in unfair, protectionist, or noncompetitive trade or investment advantages for domestic firms vis-à-vis competitors from other countries also party to the agreement. In the event that such interventions occur, these agreements make provision for penalties and retaliatory action by partner states, and also establish binding dispute-resolution mechanisms to which aggrieved parties can appeal for relief. These arrangements constrain states not only by creating markets that extend beyond the boundaries of their sovereign jurisdiction, but also by limiting their ability to use interventionist instruments *within* their jurisdiction to achieve social and economic objectives that are not being met by the interplay of private actors in markets. Once communication, culture, and information are defined as productive industries and marketable commodities, they too become subject to this logic.

As Vincent Mosco (1993, 193) points out, the FTA was "the first trade agreement in history that extends free trade in goods to the service and investment sectors, including communication" – a move that was subsequently replicated in NAFTA and GATS. These agreements apply to communication and culture in several ways, most significantly by committing signatories to extend "national treatment" to non-national enterprises. Of course, this means Canadian firms enjoy national treatment in the United States, though it is hard to imagine comparatively smaller Canadian companies posing a significant competitive threat to the more formidable enterprises already established in the US market, especially in the communication and cultural sectors. Indeed, the inability of Canadian cultural producers to compete effectively with American media titans even *within Canada* has historically provided the imperative for measures such as content regulation and a state-owned national broadcaster in this country.

Of similar concern are provisions that restrict the ability of parties to the agreement to establish state-supported institutions that will be the sole provider of goods or services (including communication) in a given territory. Given that, in the past, Canada has used Crown corporations and regulated monopolies to redress market failures by providing services where companies will not (e.g., regulated provincial telephone monopolies providing service to sparse, far-flung Prairie populations), these restrictions have potentially serious implications. While states are not expressly forbidden from establishing and owning such enterprises, Mosco (1993, 204) points out that they may do so only under

> a set of restrictions so severe as to make it unlikely that Canada will ever again be able to establish crown corporations ... The monopoly is forbidden from operating in such a fashion that permits it to be anti-competitive, nationalistic, or subsidize one of its services, like first class mail, basic telephone, or basic cable

> television services, out of revenues from other services. In essence, one is permitted to establish a crown corporation provided that it does not behave like one.

Despite the problems associated with state ownership, regulated monopolies, and cross-subsidization, had such restrictions existed over the past century, universal access to relatively affordable basic telephone service in Canada might never have been realized.

Each of the agreements being discussed here does formally exempt cultural industries, ostensibly allowing for the maintenance of protectionist market interventions in culturally sensitive areas such as broadcasting, a provision for which Canada has argued consistently and strenuously. The actual effectiveness of this formal exemption, however, is potentially quite limited relative to other provisions in these agreements. The FTA provisions on cultural industries, confirmed in NAFTA, are instructive in this regard. Article 2005(1) of the FTA states, "Cultural industries are exempt from the provisions of this agreement." Immediately following this provision, Article 2005(2) states, "Notwithstanding any other provision of the Agreement, a Party may take measures of equivalent commercial effect in response to actions that would have been inconsistent with this agreement but for paragraph 1." In other words, in the event that one country exercises its right under the agreement to protect its cultural industries, another country is permitted to take retaliatory measures *in another sector.* Thus Canada is formally permitted to protect its cultural industries only if it is willing to bear retaliatory measures in another sector, a risk that might not always be worth taking.

This is precisely what occurred in the case of Canada's efforts to protect its domestic magazine industry from competition by American magazines (Lorimer and Gasher 2001, 194-8). From its inception, the domestic magazine industry in Canada has been vulnerable to continental pressure. Established magazines with large circulations and revenue bases in the United States could move into the Canadian market with only minor additional production costs, and could sell advertising space at rates substantially lower than small-circulation

Canadian magazines needed to charge in order to survive. So, beginning in the nineteenth century, Canadian magazine publishers were granted a postal subsidy to offset their costs. In 1965 the Canadian government decided to protect domestic magazines further by allowing advertisers to claim a tax deduction for the cost of ads placed in Canadian publications (defined as 75 percent owned by Canadians), and by banning the importation of "split-run" editions (publications whose editorial content originated outside Canada but whose advertising was replaced with material from Canadian advertisers). These measures generally succeeded in securing advertising revenue and subscription bases for Canadian magazines. In 1993 US publishing giant Time-Warner accomplished an end-run around these policies by drastically reducing its advertising rates to offset the tax savings advertisers would receive for placing ads in rival Canadian publications. It then used satellite technology to beam editorial content to printing facilities in Canada, which, combined with minimal additional content, created a "Canadian edition" that was not technically a split-run. Canada responded by applying an 80 percent excise tax on all Canadian advertising sold by foreign magazines, payable by the publication's printer or distributor.

In 1997, under the auspices of the WTO, the US government challenged this excise tax, as well as Canada's postal subsidy for magazines and the ban on the importation of split-run publications, as violations of NAFTA. The WTO ruled in favour of the United States, deciding that magazines were a good, and that the postal subsidies and importation ban constituted unfair trading practices. In response, the Canadian government proposed a new scheme: rather than giving postal subsidies directly to publishers, it would grant funds to Canada Post earmarked for deposit in publishers' accounts with the corporation; it would require that only Canadian magazines be allowed to sell advertising space to firms seeking to reach Canadian readers; and it would replace the outright ban on importation with a $250,000 fine. Fearing that these modified measures might pass muster with the WTO as legitimate cultural exemptions, the United States responded by threatening massive retaliation in unrelated sectors of the Canadian economy. Unwilling to bear such consequences, which would probably also be supported by the NAFTA provisions concerning "equivalent measures" in response to the protection of cultural industries, the Canadian government backed down. A deal was struck that allowed US publishers to sell split-run magazines in Canada, allowed Canadian advertisers to deduct the cost of ads placed in US publications as a business expense, and lowered the minimum requirement for "Canadian ownership" of publications from 75 to 51 percent. A more complete concession could scarcely be imagined.

Rowland Lorimer and Mike Gasher (2001, 198) have described this as a "gutting" of Canada's long-standing, democratically enacted cultural policy regarding the national interest in a domestic magazine industry. It is important to emphasize that this took place *despite* the formal exemption enjoyed by cultural industries under NAFTA. Indeed, it occurred because – formal exemption notwithstanding – policy protecting cultural industries is still subject to severe economic retaliation under the terms of the global trade regime. Under these conditions, the actual meaning of the cultural exemption, for which the Canadian government has fought so gamely, is difficult to ascertain. Canadians have been led to believe that their right to use the Canadian state to protect domestic cultural industries against the onslaught of American mass culture is secured by the exemption of these industries from the commercial logic of free trade. But the case of the magazine industry indicates quite the opposite. As Mosco (1993, 196-7) correctly observes, far from placing culture beyond its reach, the new regime in global trade "redefines the very concept of culture to be a marketable commodity ... if you agree to permit commercial retaliation against cultural subsidies, then you have agreed to define culture as commodity." In this respect, the NAFTA exemption-that-is-not-an-exemption seems to accomplish two purposes. Superficially, it allows the Canadian government to calm the fears of its citizens regarding a loss of cultural autonomy under the regime of neoliberal globalization. Substantively, it gives effect to the long-standing US opposition to Canadian cultural protectionism, as now "for the first time the U.S. has the power of a treaty to back its claim to the right of retaliation against Canadian attempts to preserve and build on its national culture" (199).

Evidence of how neoliberalism and globalization combine to limit the ability of the Canadian state to meet nonmarket social goals in communication can also be found in the provisions in these agreements regarding liberalization of trade and competition in telecommunication services. Chapter 13 of NAFTA covers telecommunications, and it requires that parties to the agreement "ensure that persons of another party have access to and use of any public telecommunications transport network or service, including private leased circuits, offered in its territory or across its borders for the conduct of their business, on reasonable and non-discriminatory terms and conditions" (art. 1302.1). In other words, the Canadian government cannot discriminate against American firms by placing a tariff on their use of Canadian telecommunication infrastructure. Canadian firms enjoy the same nondiscriminatory access to telecommunication infrastructure in the United States. These provisions on access, replicated in the GATS Annex on Telecommunications, reflect the view that open telecommunication networks are the essential infrastructure of the global "knowledge-based" economy of the "information society." They also undermine the possibility of requiring telecommunication firms to grant priority access to domestic content. In addressing the issue of Canadian content on the Internet, the Information Highway Advisory Council recommended that the government seek to develop and enforce rules that "strengthen the principle of priority carriage for all licensed Canadian programming services on all distribution systems" (IHAC 1995, 124). Precisely this sort of regulation is expressly prohibited by the nondiscrimination provisions described above.

These agreements go further, however, and set out the terms by which telecommunication services themselves may, or may not, be regulated. In this regard, NAFTA makes a distinction between "enhanced or value-added" and "basic" telecommunication services. As defined in article 1310, "*enhanced or value-added services* means those telecommunications employing computer processing applications that: (a) act on the format, content, code, protocol or similar aspects of a customer's transmitted information; (b) provide a customer with additional, different or restructured information; or (c) involve customer interaction with stored information." Essentially, this definition covers all but the most elementary services mediated by a computer or related device. "Basic" services are left undefined but, given the expansive definition of enhanced services, "basic" cannot mean much more than plain old telephone service and the most simple forms of data transmission. GATS makes a similar distinction, but provides examples for greater detail: basic services include elementary voice and data transmission, telex, telegraph, and facsimile; enhanced services include on-line data processing, on-line database storage and retrieval (i.e., access to the World Wide Web), electronic data interchange, e-mail, and voice mail.

These distinctions set out the services that may continue to be subject to significant state regulation and those that may not. NAFTA requires that parties refrain from discriminating in favour of domestic firms in licensing the provision of enhanced or value-added telecommunication services (art. 1301.1). It also stipulates, "No party may require a person providing enhanced or value-added services to: (a) provide those services to the public generally; (b) cost justify its rates; (c) file a tariff; (d) interconnect its networks with any particular customer or network; or (e) conform with any particular standard or technical regulation for interconnection other than for interconnection to a public telecommunications transport network" (art. 1303.2). Though GATS provisions regarding telecommunication regulation differ slightly from those in NAFTA, the result is similar: when it comes to so-called enhanced or value-added telecommunication services such as access to the Internet, the Canadian state has contracted away its ability to enforce public service goals and obligations (such as, for example, universal service at reasonable rates) upon private enterprises (including non-Canadian firms) profiting from the provision of these services, by agreeing not to use regulatory instruments aimed at such outcomes. This move entails serious democratic liabilities. As

Winseck (1998, 226-7) observes, the new regime "removes the new 'technologies of freedom' from the political/public policy agenda. Defining basic communication services very narrowly, the new trade-in-services regimes prevent media policies from expanding the range of publicly regulated services commensurate with the emerging technologies and felt needs of some citizens as we enter the 'information age' ... The new global trade regimes simply exist on expanding the information commodity rather than electronic public spheres of communication."

Significantly, provisions liberalizing markets in telecommunication services could also further undermine the already tenuous exemption of cultural industries from these agreements. As the technological convergence of various media forms into digital networks proceeds, much cultural content will increasingly be mediated by services that can be defined under this regime as "enhanced" or "value-added" and, therefore, protected from significant state regulation. In this manner, "the enhanced services provisions offer a convenient way of escaping Canadian media and cultural policies" (Winseck 1998, 311).

The spirit of market expansion, privatization, and regulatory liberalization enshrined in the various trade agreements that make up the formal constitution of capitalist globalization has also animated the imagination and development of the so-called Global Information Infrastructure (GII). At a private meeting in 1995 of leaders of the Group of Seven industrialized countries in Brussels, the United States set forth a plan for the elaboration of the digital infrastructure of global capitalism and electronic commerce. As described by Marc Raboy (2002, 127), the GII plan represents something of a culmination of the logic of globalization as it has been discussed here: "Largely written in Washington, the plan to establish a 'Global Information Infrastructure,' adopted by the G7 at that meeting, represented an imperial triumph of unprecedented scope. It enshrined a single vision, program and policy framework for the role of communication technology as a means of achieving an idealized global society driven by the market forces of transnational capital."

Such a project would require, at a minimum, an agreement among the world's most powerful countries on a mode of governance in relation to communication and technology. This mode of governance would be one in which transnational private enterprise would be allowed to take the lead and reap the rewards, and the role of the state would be primarily to maintain conditions conducive to investment, innovation, commercial development, and technological progress. This is precisely the approach confirmed in the G8's Okinawa Charter on the Global Information Society, adopted in 2000 "with the global business agenda in mind" (Raboy 2001, 151).

In light of the above, domestic policies such as those discussed in Chapter 2 come into sharper focus. Take, for example, the CRTC's 1995 "convergence" decision to allow cross-platform concentration of ownership in the name of competitiveness, and its categorical declaration in 1999 that it would not regulate the Internet. Given Canada's commitments and constraints under the regime described above, the CRTC could not have decided otherwise. The same might be said of the IHAC recommendations that the development of ICTs in Canada proceed under the auspices of privatization, market principles, and liberalized (i.e., lightly regulated) competition. Again, in the context of the global trade regime to which Canada has committed itself, such policy decisions must be regarded as little more than foregone conclusions. These facts radically undermine claims that the (limited) consultation processes preceding the CRTC's reports on convergence and new media, and IHAC's two final reports, were substantially democratic. Processes in which certain policy options are effectively invalidated from the outset cannot be democratic, because their outcomes cannot be responsive to those who might advocate positions that are prejudicially discounted, even if they are persuasive or represent majority views. In this instance, consultation is reduced to a public relations exercise, instrumental to a cynical legitimation of decisions that have already been made and that have, in fact, been exempted from democratic consideration.

This is just one way in which the encounter of Canadian communication with capitalist globalization in Canadian communication can be viewed critically in democratic terms. The economic arrangements undergirding the globalization of Canadian communication policy are deeply political because they set the terms according to which Canadians may distribute resources and organize their collective lives through the application of state authority. In this sense, these arrangements compose what has been described as "a new constitution for North America" (Mosco 1993, 194). Furthermore, communication and culture, so deeply implicated in this new regime, are also linked inextricably to democratic citizenship, insofar as "cultural development can be defined as the process by which human beings acquire the individual and collective resources necessary to participate in public life" (Raboy et al. 1994, 2). Given this, one might expect that the processes by which these arrangements were constructed would be substantially democratic, allowing opportunities for participation by citizens, inclusive of a diversity of perspectives, and responsive to the public interest. One might also expect that the outcomes of these processes would be sensitive to the place of culture and communication in meaningful democratic citizenship.

Sadly, the Canadian approach to the globalization of communication has failed on both counts. The processes by which Canada has given the consent of its citizens to the regime of capitalist globalization – primarily high-level negotiations surrounding trade and investment agreements – have been marked by lack of transparency, the scarcity of opportunities for effective citizen participation, exclusivity, and overrepresentation of the interests of powerful private corporations. Despite their clear implications for communication policy, and despite a tradition of significant public engagement in this policy area, the negotiations surrounding the FTA and NAFTA "occurred with no hint of the traditional public consultation process ... and radically shifted the intent and process of communication governance" (Abramson and Raboy 1999, 786-7). Conversely, major corporate players in the communication sector have been well represented in these forums (Mosco 1993, 206). Globalization, in this sense, is "undoing the democratization" that once provided at least some degree of transparency, accountability, and nominal participation in Canada's communication policy regime (Raboy 2002, 111). At the 1995 G7 meetings, from which emerged the plan for a GII that would materialize the global business elite's fondest hopes for the policy regime surrounding ICTs, representatives of several major transnational media, technology, and communication corporations were granted official status. Meanwhile, transnational civil society groups were "relegated to the margins of unofficial intervention" (128).

Clearly, as Winseck (1998, 217) reports, "support for the 'globalization' of domestic telecoms policy was widely diffused throughout the power centers of Canada." It is not clear that the same consensus existed among all those who were concerned about the public interest in communication and culture, but who may not have been situated at the centre of political and economic power in Canada. If differences of viewpoint did exist on these questions – as we might expect in a diverse society such as Canada, one with a strong tradition of public service norms and state intervention in communication markets – these were denied a hearing in institutional venues connected to effective decision making. The practice of global communication policy making in recent decades thus mirrors exactly the exclusive and nonparticipatory (at least as far as everyday citizens are concerned) nature of Canadian domestic policy making surrounding ICTs discussed in Chapter 2.

Under almost any credible definition of democracy, the character of these processes is troubling. This is especially so given the political stakes involved in outcomes that will determine how, and to what ends, ICTs are elaborated in our midst. As Raboy (2002, 129) writes, under the current regime, the development of these technologies proceeds as "an imperial project with enormous implications for the future of democracy and human rights, insofar as it is based on political decision-making at a level where there is no accountability, on the recognized autonomy of private capital, and on the formal exclusion of the institutions of civil society." This model leaves citizens with the streets as a venue in which to gather and voice their concerns, though, in recent antiglobalization protests in Canada and throughout the world, even these erstwhile public spaces have become subject to increasingly strict, and sometimes violent, private and state control.

The outcomes of these undemocratic processes serve to further undermine the prospects of a democratic elaboration of ICTs. Canada's participation in constituting the regime of capitalist globalization has certainly achieved the goal of securing access to global markets for powerful Canadian corporations, but the cost in terms of diminished capacity for autonomous domestic policy making may be too high for a democracy to bear. This capacity is diminished not because Canada must cooperate with other nations in crafting a regulatory and policy regime for global communication technologies. Mass communication has always been an issue that crosses borders. It is diminished because, in agreeing to defer these functions to the interplay of powerful private actors in markets that are free of constraint and protected from redistributive intervention, Canada and other capitalist countries effectively remove issues surrounding these technologies from public, political consideration. Under the contemporary regime, the essentially political nature of communication, information, and culture is obscured, as each is reduced to the status of a commodity. As commodities, they are functionally depoliticized: subsumed under binding agreements that limit the ability of publicly constituted, democratic political authorities to direct their development and distribution. In a democracy, determinations regarding common goods are made by public citizens participating in inclusive and responsive political processes. The regime described here substantially deprives Canadians of meaningful citizenship opportunities in relation to the public goods of communication, culture, and technology. Instead, they are left to confront the might of transnational capital primarily as isolated, private consumers, wielding only the power of their individual commercial choices. Sadly, the new global regime in communication policy thus works to minimize democratic participation in cultural and political life and decision making, even as new communication technologies themselves raise the possibility of something approaching a transnational, democratic public sphere.

It is tempting to portray capitalist globalization as an abstract, external, irresistible force imposing itself upon the Canadian state and its citizens against their will. This view is not entirely without substance. The vision for a global, privatized, and market-driven information infrastructure presented by the United States to the G7 in 1995 was presented as a plan "to be taken home and put into place by national policy makers" – a direction that the Canadian government fairly dutifully obeyed through the course of the subsequent CRTC and IHAC policy exercises (Abramson and Raboy 1999, 781). Canada's consistent struggle to maintain cultural exemptions (despite their questionable force) against equally consistent efforts by the United States and WTO to undermine and delegitimize these also suggests a certain resistance by the Canadian state to relinquish completely the reins to global markets and the private interests that rule them. Nevertheless, the Canadian state was a willing and enthusiastic participant in the construction of the economic and political regime of capitalist globalization. The Canadian state's choice to pursue the development of new information and communication technology along the paths of privatization, concentrated ownership, and regulatory liberalization, and to exempt these issues from democratic political consideration, was not made under duress.

These dynamics began in Canada even before agreements such as the FTA and NAFTA were signed. For example, the CRTC adopted a US-inspired distinction between basic and enhanced services that effectively placed emerging digital services such as Internet access and e-mail "beyond the reach of public service obligations" in 1984, well before such provisions were enshrined in the FTA/NAFTA/GATS cycle of agreements (Winseck 1998, 219). This, with other examples, suggests that Canada was not forced to commit itself to market liberalization in communication by virtue of signing international trade agreements, but rather that Canada consented to these agreements because it was already committed to market liberalization. The same might be said of the reversal of the Canadian tradition of democratic consultation in communication policy making. Policy surrounding ICTs has not been withdrawn from relatively inclusive, participatory, and responsive processes because of globalization. Rather, the deeply undemocratic character of Canada's approach to the globalization of communication is an extension of its recently undemocratic domestic communication policy making regime. When it comes to bringing democratic processes to bear in determining how the public goods of new ICTs will be developed and deployed in Canada, we have seen the enemy, and it is not an abstract "other" called globalization. It is, more accurately speaking, us.

## Democratizing Globalization

Along with the tendency to portray the undemocratic aspects of economic globalization as imposed from without by an abstract and alien force, there is a tendency to concede that the pathologies of economic globalization are intractable, and that effective democracy is irretrievable under these conditions. Such a concession, of course, amounts to a guarantee that Canada's encounter with globalization in general, and with ICTs in particular, will continue along an undemocratic trajectory. To recover democracy in light of what has been discussed above requires, at a minimum, that citizens continue to have the courage to imagine its possibility despite its elusiveness.

Globalization is not an iron law of history. That being said, the energies driving the current episode of economic globalization seem unlikely to abate any time soon. The question is whether and how democratic political control over areas such as communication, culture, and technological development might be retrieved from the logic of commodification and transnational marketization that dominates the present moment. One possibility is that transnational civil society movements, themselves made possib116
le by global communication networks, will succeed in their demands for a democratization of the international institutions governing the global economy. As Cameron and Stein (2002a, 154-7) point out, hope for this outcome should be measured at best:

> Currently, citizens and networks are not subjected to anything like the standards of accountability, transparency, and representativeness imposed on democratic governments, despite all the legitimate criticisms that can be made of these state-based processes .... Although groups of citizens are mobilizing to hold institutions accountable and to increase transparency, at present the accountability of international institutions is at best embryonic. International institutions remain a poor alternative to democratic, legitimate and accountable states.

In their view, "The most promising arena for rule-governed popular contestation remains the democratic state [because] the modern, rule-governed democratic state is still unmatched in its capacity to provide accountability and representation" (Cameron and Stein 2002a, 157). It is also the only agent able (and, it could be argued, designed) to confront powerful private actors and compel them to submit to the redistribution of social resources according to norms of social justice and the common good. States do not always perform this role, and they can often undermine it. Nevertheless, democracy would seem to begin at home. Specifically, it resides in the democratic fortunes of the national state.

The trick, of course, is to compel state institutions to make good on their formal commitment to democracy in both policy processes and outcomes. This will be difficult so long as states such as Canada remain wedded to a conception of globalization that is indistinguishable from the liberation of capitalist enterprise in transnational markets, a conception that entails the depoliticization of crucial social goods such as communication and culture, and effectively delivers the power to determine these issues to major private interests in the economy. Therefore the prospects of democratizing communication, culture, and technology under contemporary conditions hinge decisively on the possibility of breaking the link between neoliberalism and globalization (Raboy 2002). Surrendered wholly to the imperatives of liberated transnational markets and the private interests that dominate them, neither the processes nor the outcomes surrounding the development of ICTs are likely to be substantially democratic.

Another way of putting the matter is to say that the Canadian state must be compelled to alter its perception of its role in response to globalization. To this point, the Canadian state has configured itself as a "handmaiden state," whose purpose it is to adapt its domestic regime as required by the imperatives of economic competitiveness in a global economy (Cameron and Stein 2002a, 148). It achieves this by enforcing measures perceived as conducive to profitable investment and enterprise, including privatizing social goods and public resources, minimizing regulatory intervention in markets, enhancing the competitiveness of so-called national champions in global markets, preparing a workforce for productive employment under dynamic conditions, and actively sponsoring technological innovation and development. This fairly summarizes the Canadian state's approach to issues surrounding ICTs in recent decades. This approach articulates well with the tradition of technological nationalism in Canada, a form of nationalism that historically has served the interests of industrial capitalism and continental cultural assimilation more directly than it has served Canadian democracy. It is also an approach in which, as Cameron and Stein (2002a, 148) describe, "the space for political and economic policy choices shrinks, and state capacity to make choices in the space that remains declines." Clearly, such a response bodes ill for democracy.

An alternative response would be for the Canadian state to work actively to redress the democratic failings of globalization. This does not require Canada to pretend global capitalism does not exist, or simply to absent itself from its dynamics. As Raboy (2002, 120-1) has perceptively written, "For Canada, seeking to distinguish itself from the United States has always been not so much a reflex of withdrawal but part of an attempt toward openness to the rest of the world on its own terms." Even within the context of globalization, the Canadian state could strive to provide a platform upon which Canadians could express, and act upon, their own democratically determined preferences for living in the world constructed by new ICTs. This does not necessarily mean the enforcement of a state-led cultural nationalism. It does necessarily mean opening domestic policy process to real democratic engagement, and promoting an extension of democratic norms from the domestic to the transnational level. To be adequately democratic, these processes would have to feature something like the following:

- *publicity* and *transparency* (as opposed to the privacy and secrecy of closed meetings and corporate boardrooms)
- ample *opportunities*, and material support, for formal participation by citizens as equals
- active *inclusion* of a plurality of interests, constituencies, and policy options
- outcomes that are genuinely *responsive* to the public good as expressed through these processes.

Such commitments would be a marked departure from the undemocratic spirit that has lately attended policy making regarding ICTs in Canada and internationally. They would, however, be consistent with the more inclusive, participatory, and responsive tradition that reflects the best face of Canadian democratic nationalism and has been, until recently, a distinctively Canadian norm in this area. Recovering this tradition domestically would require little vision, but considerable will; extending it into the arenas of globalization would require courage and stamina, for what has been called "the long march through the institutions ... tied to the broader project of the democratization of global governance" (Raboy 2002, 135). At a minimum, this would involve Canada in active support of multilateral initiatives already engaged in democratic reform (such as the efforts of UNESCO and the International Telecommunication Union). It would also require Canada to take a vanguard position in demanding democratic reform of the exclusive, undemocratic institutions – the G7/G8, the OECD, the WTO – currently dominating the policy regime of capitalist globalization.

Such a democratization of the politics of communication, technology, and culture would require more than procedural change. It would also require a radical shift in our basic understanding of communication and culture as social goods. As Raboy (2002, 134) writes, "Credibility will need to be given to the idea that the global communications environment, from the conventional airwaves to outer space, is a public resource, to be organized, managed and regulated in the global public interest." Thus, democratization requires that ICTs, and the practices they mediate, cease to be regarded solely as commodities for trade between self-regarding actors in global markets shorn of state regulation. Instead, they must be conceived as common goods and social practices that are foundational to cultural autonomy and democratic citizenship.

As such, the development and distribution of ICTs cannot be left wholly to markets constructed for private economic accumulation, and characterized by marked inequalities of power and outcome. Rather, justice requires that the demands of the public good, as determined through genuinely democratic processes, be enforced upon the market distribution of communication and cultural resources. If we are concerned for democratic outcomes surrounding ICTs, market logic and the actors that command it cannot be allowed to reign in this sector as exclusively as they do at present. Once again, Canada is well positioned to make this case, given its historical sensitivity to the vulnerability of self-determined communication and cultural sovereignty to market forces, and its historical willingness to use state instruments to intervene in markets to secure public interest outcomes.

To be sure, Canada has been persistent in its efforts to secure the exemption of culture from the global regime of trade in goods and services, and has often played a leadership role in marshalling other nations to this cause (Raboy 2002, 130-1). For example, in 1999 the Cultural Industries Sectoral Advisory Group on International Trade sponsored by the Departments of Canadian Heritage and Foreign Affairs and International Trade issued a report on the DFAIT website calling not only for continued efforts to exempt culture from international trade agreements, but also for the development of a "new international instrument on cultural diversity." Such an instrument would "acknowledge the legitimate role of domestic cultural policies in ensuring cultural diversity" by asserting the distinctiveness of indigenous cultural goods and services and the policies necessary to secure domestic access to these. The instrument would then "set out rules on the kind of domestic regulatory and other measures that countries can and cannot use to enhance cultural and linguistic diversity; and establish how trade disciplines would apply or not apply to cultural measures that meet the agreed-upon rules." Subsequently, Canada has advocated this position in international fora such as UNESCO and the International Network on Cultural Policy (Raboy 2002, 130-1).

These efforts are commendable and significant, but they reflect a nagging ambivalence in Canada's position on capitalist globalization and communication, rather than a consistent and coherent resolve to secure national cultural autonomy and a democratic political space. Canada's support for the development of ICTs under the auspices of a global, market-based regime that serves the interests of capitalist industry in this country has been energetic and unambiguous. Given the clarity and consistency of this overall position, it is hard to imagine the Canadian state going to the wall over its sincere desire to exempt "culture" from a market logic whose application to information, communication, and technology it otherwise endorses so strenuously. Securing an autonomous, self-determined, and democratic encounter for Canadians with ICTs in the age of globalization will require a modification of market principles that extends beyond rearguard campaigns for a feeble cultural exemption. Such modification would undoubtedly come with a price that might include diminished access to foreign markets and investment, retaliation by trade partners, and reduced technological dynamism. The questions are whether Canadians would be willing to pay this price for democratic control of their communication environment and, if so, how they might impose this choice upon those who govern them.

## Chapter 3

### Strengths

- Canada has a long history of trying to balance cultural sovereignty and industrial competitiveness in communication media.
- Canada has led the effort to establish an international instrument for the exemption of cultural industries from antiprotectionist trade agreements.
- New ICTs can provide a means for Canadian citizens to participate in global civil society movements and transnational public spheres.

### Weaknesses

- New ICTs are tied to the challenges posed by globalization for national political autonomy and state intervention in markets.
- The protection of culture and communication has been made more difficult by their redefinition as commodities in global markets.
- Decision making about the global economy, including the regulation and development of ICTs, takes place primarily in venues that lack transparency, adequate citizen representation, and democratic accountability.

# 4 TECHNOLOGIES OF POLITICAL COMMUNICATION IN CANADA

Writing in 1958, Raymond Williams (1983, 315) sounds as though he is describing our present political scene, in which the casting of bricks at plate-glass windows has come to be viewed by many as a more authentic political act than the casting of votes: "If people cannot have official democracy, they will have unofficial democracy, in any of its possible forms, from the armed revolt or riot ... to the quietest but most alarming form – a general sullenness and withdrawal of interest." At the turn of the millennium, Canada seems to be suffering from an acute case of Williams's "unofficial democracy": ours is a political condition in which violent state repression of popular demonstrations on the one hand, and a climate of widespread disaffection and disengagement on the other, are so routine they do not astonish us. According to Williams, "These characteristic marks of our civilization are symptoms of a basic failure in *communication*" (315, emphasis added). He goes on to suggest that the solution lies in "adopting a different attitude to [communication], one which will ensure that its origins are genuinely multiple, and that all the sources have access to the common channels" (316).

That prescription sounds a lot like the Internet, or at least its idealized version, as presented by those who associate its particular properties – speed and reach, decentralized architecture, potential for interactive engagement, immediate and widespread publication, access to voluminous information – with a democratic renaissance. Surely, according to the mythology surrounding the Internet, a technology that does all this can only complement democracy, *and maybe even save it.*

Have the Internet, and its related technologies, contributed to the realization of enhanced democratic political communication? Has the reorientation of political communication around digital communication media resulted in political practices, institutions, and relationships that can be described as more inclusive, participatory, and responsive than those that characterized previous media environments? These questions are the subject of this chapter. Their answers reside, at least in part, in the actual uses to which Canadian political actors and institutions put these technologies in their everyday practices. Few serious political actors and institutions today, whether mainstream or marginal, have failed to avail themselves of, or are unaffected by, the utilities of digital ICTs. Indeed, most have embraced these technologies enthusiastically. This chapter focuses on four categories of actors and institutions that occupy the contemporary political landscape in Canada – government, political parties, advocacy groups and social movements, and individual citizens – and inquires into the characteristics of their use of new ICTs.

## Government

Governments throughout Canada have been intimately involved in the development and deployment of ICTs, perhaps in response to the Information Highway Advisory Council's repeated admonitions that government must become a "model user" of these technologies. In fact, IHAC felt so strongly about this that it reiterated this recommendation no fewer that three times in its initial final report: "The government should be a role model in the cost-effective use and promotion of information technology" (IHAC 1995, rec. 9.11); "The government will make it a priority to become a world leader in the rapid introduction and generalized use of electronic information and communication systems, and in affording all Canadians the opportunity to communicate and interact electronically with its departments and agencies in either official language" (rec. 15.1); and, for emphasis, "The government must be a role model in the timely deployment of the information highway and in the cost-effective use and promotion of information and communication technology" (rec. 15.2).

The Canadian government has taken this advice to heart and acted upon it in a decisive, deliberate, and sustained manner, throwing itself enthusiastically into the information "revolution" (Alexander 2000). The government's use of new ICTs has taken many forms. In this section, I will concentrate on two categories into which some of these activities can be gathered: e-government and e-democracy. E-government refers to the use of new ICTs in the operations of government and the delivery of government services. E-democracy refers to the use of these technologies as a medium for citizen engagement in government policy processes and decision making. I will examine each in turn.

## E-Government

In 2003, for the third year running, the Canadian government was ranked number one in the world for e-government service delivery in an influential annual study conducted by an international consulting company. Canada topped the list in terms of "e-Government maturity," as the only country to have reached the "fifth plateau" of service breadth and sophistication on the basis of its "overall service transformation" (Accenture 2003). This recognition was undoubtedly pleasing for the Canadian government, which, in 1999, announced from the throne its intention to become "known around the world as the government most connected to its citizens, with Canadians able to access all government information and services on-line at the time and place of their choosing" (Canada 1999). This commitment had, in fact, been a matter of policy prior to 1999. As early as 1993, the Treasury Board Secretariat had signalled its intention to "renew" government services using a variety of information technology applications (Treasury Board Secretariat 1993). The project materialized decisively in the Government On-Line (GOL) initiative, announced in 1999, with an initial budget of $160 million and total direct expenditures projected to reach $880 million by 2006 (Treasury Board Secretariat 2003, app. A). This figure represents only a fraction of the total cost of going digital. In 2001 the government estimated that the total cost of placing all key services on-line could reach up to $2 billion, and suggested these additional costs could be funded by departments and agencies via internal cost savings and resource reallocation (Office of the Auditor General 2003, 1-2).

The GOL initiative has proceeded along two related paths. The first has been a comprehensive effort to make government services and information available via the Internet, to the point of establishing electronic service delivery as the primary mode of contact between citizens and government. In a combined effort managed by the Treasury Board Secretariat, the Privy Council Office, and the Ministry of Public Works and Government Services, the Canadian government has established a formidable on-line presence in a relatively short period of time. As of 2000, all government departments and agencies had websites and the Canada site Internet portal was established. By 2001 nine services were completely on-line. This number increased to forty-five services by 2003 and, in its 2004 annual report, the GOL initiative projected that by 2005 all of the federal government's "most commonly used services" will be available on-line, a goal that includes over 130 services from thirty federal departments and agencies (Public Works and Government Services 2004, 6, 14). These services include employment and immigration services, tax return filing, business and consumer information and program services, old age security and pension benefit applications and services, a wide variety of funding and licensing applications, and a significant range of health-related services.

The GOL initiative's electronic service delivery project has been animated by a number of priorities: the desire to configure GOL so as to realize efficiencies and cost savings in the delivery of government services; the development of quality control regimes, in the form of best practices and a common look and feel for all government websites; the effective management of information privacy and infrastructure security concerns; and an effort to make electronic services accessible, convenient, and responsive to the needs and preferences of users. This last emphasis has entailed the development of portals and gateways that group access to services and information either according to subject areas or to the profile of the users. Thus, the Canada site (canada.gc.ca) now provides fifteen subject gateways for topics including Consumer Information; Culture, Heritage and Recreation; Rural and Remote Services; Justice and the Law; and Jobs, Workers, Training and Careers. It also provides nine "audience"-based gateways for profiles such as Aboriginal Peoples; Canadian Business; Newcomers to Canada; Seniors; Youth; and Persons with Disabilities. The 2004 GOL report documented 16 million visits to the Canada site, a 21 percent increase since 2002 (Public Works and Government Services 2004, 10).

This emphasis on a client-centred approach to electronic service delivery entails "locating" access to ranges of services and information that might derive from different programs, agencies, departments, and even levels of government. The promise of more effective and efficient service delivery and consumption via electronic media is premised on a fairly simple idea: instead of forcing a person seeking a basket of related services to physically travel to five different offices to fill out ten slightly different forms, or to make a dozen frustrating telephone calls, why not enable her to find them by electronically visiting a single site that acts as a platform for "one-stop shopping"? This allows consumers of government services to access a broad range of information, agencies, and services from a single remote location (whether it is their own home, their workplace, or a public terminal in their community) in a manner that is generally conceded to be much more efficient than the traditional gauntlet of telephone calls, queues in multiple offices, and postal correspondence. Implicit in this approach is the understanding that the services or information a given person may wish to access are often dispensed by multiple agencies, sometimes spanning federal, provincial, and municipal jurisdictions. Therefore ensuring a "seamless service experience" requires an "integrated service delivery network" that provides click-of-the-mouse access to services and information not just *within*, but *across* the organizational divisions of government (Treasury Board Secretariat 2003, 3, 28).

Cultivation of this whole-of-government approach represents the second path of the GOL initiative. As anyone remotely familiar with the politics of public administration would attest, this is an undertaking of extreme complexity, requiring a balance of centralized coordination and standards with decentralized, multisectoral, and cross-jurisdictional collaboration, a veritable culture shift within the Canadian bureaucracy. The project is made even more complex by the recent trend toward privatization and devolution of many government services through public-private partnership arrangements in which private sector agencies assume the role of service providers. The whole-of-government approach to electronic service delivery must, in this respect, also encompass a range of nongovernmental partners. At one level, the GOL initiative has responded to this complexity by working to develop an electronic architecture capable of supporting the integration of multiple service and information delivery channels, an architecture in which there is "no wrong door" to electronic access (Treasury Board Secretariat 2003, 9).

More generally, however, these efforts are of a piece with the broader organizational move away from governance organized on a departmental model, and toward governance based on a networking model (Alcock and Lenihan 2001, 19). In relation to governance, departments are like silos: the material they contain is closely defined and limited, they are organized vertically, and their emphasis is on procedural consistency. Their walls are relatively impermeable, meaning inflows and outflows (of information, expertise, resources, decision making, services, etc.) enter and exit only at the top and bottom. They operate more or less independently of other silos, and in relatively fixed positions with respect to them. Thus we have, in Canada, a Department of Agriculture, a Department of the Environment, a Customs and Revenue Agency, a Food Inspection Agency, and so on. In the departmental model of governance, each of these is a silo of sorts, responsible for the development of policy and programs, and the delivery of services, within its closely defined jurisdiction. Networks differ from departments in significant ways: they are loosely defined and have fluid limits; they cohere around projects defined by shared concern and responsibility for particular outcomes; they are horizontally distributed; they are permeable, providing multiple points of access and egress; they mesh easily with other networks; and they are dynamic in their membership and construction.

The organization of governance on the departmental model, characteristic of most modern bureaucratic states, has been at least partly a function of the realities and limitations of information and document management in paper-based systems. Any model of governance relies heavily on communication and the movement of information and documents. When information moves slowly and communication takes time, the values of efficiency, coordination, and scrutiny recommend a departmental structure. When information and documents can move quickly and communication is potentially instant, however, other models become possible without sacrificing efficiency, coordination, and scrutiny. Indeed, under such conditions – the conditions made possible by ICTs – the rigidity, hierarchy, and proceduralism of departmental organization can sometimes pose barriers to the efficient and effective execution of governance on a more distributed model.

Canada has yet to relinquish entirely the departmental model of governance for the networking model. That being said, significant energy is being devoted to the imagination and pursuit of this transition, largely in response to the deployment of ICTs. According to the Centre for Collaborative Government, "As [ICTs] proliferate, they are creating a new kind of infrastructure, one that conforms less and less to existing government boundaries ... countries such as Canada may be passing through a threshold where the government's centre of gravity is shifting from the old departmental model to a new networking model" (Alcock and Lenihan 2001, 19). The centre also reports a dawning recognition that "seamless government would require much higher levels of coordination and collaboration between departments, governments, and other service providers in the private and voluntary sectors than now exist, resulting in major increases in interdependence at every level of government activity" (Lenihan 2002a, 11).

Recognition, however, is a long way from realization. Such an architecture of intra- and intergovernmental interdependence and networking could have serious consequences, not the least of which might include a reorientation of the regional identities upon which the federal system in Canada is based, an erosion of the jurisdictional boundaries that distinguish provincial and territorial governments from the federal government, and a blurring of the lines of accountability for governmental action and inaction (Gibbins 2000). According to Roger Gibbins, "Federalism and the jurisdictional walls it erects around territorially defined communities make less sense in the context of a global village knit together by ICTs" (674). If both regional and national political identification erodes in this manner (giving way to greater identification with local and global communities, respectively) then the historical logic of the federal distribution of powers may also be undermined. And when citizens (many of whom already have difficulty sorting out which level of government does what) begin accessing integrated services through a unitary window, they may find it even more difficult to determine which government is responsible when they are dissatisfied. Such possibilities remain speculative at this point, but they do suggest that the implications of "seamless government" are not just technical and administrative, but political as well.

Yet another set of implications of the networking model accrues to the possibilities – and perhaps imperatives – it presents for the increased engagement of citizens as active agents in the governance process, as opposed to passive consumers of government services. If the transition to networked governance represents the second path of e-government, this prospect of enhanced citizen engagement mediated by new ICTs represents the third. On this path, e-government gestures in the direction of e-democracy.

## E-democracy

The idea of a more distributed, decentralized structure of governance mediated by ICTs is not necessarily limited in its potential application to service delivery. It also raises the prospect of increased citizen involvement in the formation, development, and realization of government policy, and in decision making. This is the promise of e-democracy: that ICTs make possible levels and forms of citizen engagement that have heretofore been impossible because of the obstacles the country's scale poses for the widespread information distribution and sustained communication necessary for such engagement. Technologies like the Internet, at least in theory, enable large numbers of citizens dispersed across the country to gain access to great volumes of politically relevant information, and to play a more direct and ongoing role in processes once reserved for representative or bureaucratic decision makers. The possibilities here range from posting government and policy information on-line, to use of the Internet to conduct opinion polls, surveys, and referenda, to on-line consultations with specific stakeholders, to the use of e-mail as a mode of communication with representatives, to electronically mediated deliberative and decision-making forums.

Like several other governments around the world, Canada's has recently been flirting with the possibilities of e-democracy as a complement to its more vigorous efforts in the area of electronic service delivery (Coleman and Gøtze 2001; Van Rooy 2000). Perhaps its most decisive, and arguably most successful, initiative has been a massive project to make government and policy information available via the Internet, an effort aimed at increasing citizens' awareness and building citizens' capacities around public policy issues. In most accounts, information access is also identified with more accessible and transparent governance. Governments in Canada now routinely make reams of politically significant and useful information available by electronic means, primarily the World Wide Web. Federal government departments, agencies, committees, programs, boards, commissions, task forces, councils, secretariats, and so forth maintain a formidable on-line presence. This presence largely takes the form of documentary information accessible to the general public, including a broad range of reports, transcripts and proceedings, statutes and regulations, notices, press releases, statistics, official policy statements, research documents, and program materials, as well as material published specifically for distribution via the web. Government information not published electronically is often nevertheless available via e-mail request. Most, though not all, information is available at no cost to those able to access it on-line. Locating on-line information that is relevant to one's specific purposes is not always perfectly easy, but it would be difficult to contest the fact that the remote, twenty-four-hour search, retrieval, and storage capacities of computer networks and digital interfaces have made more government information more readily available to more people than was possible prior to the existence of this medium. Research conducted under the GOL initiative reports that a "majority" of Canadians list "to obtain information" as their main reason for visiting federal websites (Public Works and Government Services 2004, 19).

Still, increased access to information, even when actively pursued by citizens, does not constitute democratic engagement. Broadly construed, democratic engagement also entails a flow of information from citizens into the policy development and decision-making processes that are the core of modern governance. At a minimum, such engagement is a condition of the legitimacy of representative governments.

More fundamentally, meaningful citizen engagement in policy development and decision making is the substantive foundation upon which a polity's claim to a democratic identity ultimately rests. Significantly, therefore, in the contemporary political imagination, digital technologies are largely defined by the perception that they are suited to mediate forms of citizen engagement that are more robust than has been typical of liberal democracies in the late twentieth and early twenty-first centuries (Becker and Slaton 2000). Depending upon the account, this prospect holds out the possibility of bolstering the legitimacy of representative systems or, more radically, of transforming the practice of government toward more direct forms of democracy.

The Canadian government's efforts in relation to electronically mediated forms of citizen engagement in policy development and decision making have been more limited, and more ambiguous, than its decisive embrace of electronic service and information delivery. In 2000 an amendment to the Canada Elections Act authorized the chief electoral officer to "carry out studies and tests on alternative voting means, including electronic voting processes" (s. 18.1). In fact, as John Courtney (2004, 120-4) discusses in his volume on elections in this series, Elections Canada had commissioned a report on various forms of electronically mediated voting in 1998. The report suggested that three technologies – the Internet, electronic kiosks, and the telephone – "offer the prospect of significantly improving both the accessibility and efficiency of the electoral process in Canada" (KPMG 1998, 5). It also pointed out several obstacles standing in the way of deploying such systems – cost, security, privacy, public confidence – but, in general, concluded that the possible benefits merited efforts to overcome these obstacles.

Introducing the results of the 1998 study to the public in 2000, Chief Electoral Officer Jean-Pierre Kingsley stressed the need to "develop greater expertise with technology if we want the electoral process to remain relevant to young Canadians," and pointed out that "Elections Canada has now computerized virtually all of its functions, *except* the act of voting and the counting of votes" (Kingsley 2000, 1, emphasis added). Here Kingsley refers to the establishment of the National Register of Electors, the use of digital geographic information systems in the production of electoral maps and polling district assignments, and the availability of electoral information and results on-line. Nevertheless, the government's movement in the direction of electronically mediated voting has been negligible.

In 2002 Elections Canada published the results of a second study supporting the feasibility of on-line voter registration but, at the same time, stressed that issues of system security, secrecy, organizational capacity, accessibility, and voter capacity remained to be resolved before any broader implementation of electronic voting schemes could even be contemplated (Guérin and Akbar 2003, 30). In any case, electronic voting would not necessarily have a discernible positive impact on voter turnout, even among young voters. As Courtney has pointed out, those most likely to use the Internet – the educated and affluent, both young and old – are also those who are already most inclined and likely to vote anyway. Thus, he concludes that "Internet voting may increase the pool of participants only marginally, if at all ... voting on the Internet could amount to little more than an additional way of casting a ballot for those who already vote and do little to address the more fundamental problem of how to increase the level of voter turnout of young Canadians" (Courtney 2004, 124).

It could also be argued that focusing on the prospect of electronic voting misses the point entirely, given that the measure of a democracy is the extent and character of the participation it allows *between,* rather than just during, elections. Signalling individual preferences through voting for particular candidates or parties is a minimalist form of democratic expression. The hope for new technologies is precisely that they might facilitate progress beyond this minimum by mediating more deliberative, dialogic forms of participation on an ongoing basis. Have Canadian governments taken seriously the potential of these technologies to mediate more vigorous forms of democratic participation?

While there is evidence to suggest that subnational levels of government – in particular, municipalities – have been quite creative in their experimentation with digitally mediated citizen engagement (Culver 2003; Lenihan 2002b), the approach taken at the federal level in Canada has been more limited. Indeed, notwithstanding the ease with which government agencies invoke the rhetoric of electronically enhanced democratic renewal in their efforts to imagine Canada as an "information" or "knowledge society," concrete efforts to make good on this rhetoric have been relatively sporadic. In January 2003 the Department of Foreign Affairs and International Trade (DFAIT) launched a program of on-line consultation in conjunction with its broader Dialogue on Foreign Policy. From launch to final report, the consultation project lasted six months, and included a series of fifteen town hall meetings, nineteen expert round-tables, meetings with provincial and territorial governments, hearings with parliamentarians, and a national forum for youth, all guided by a dialogue paper issued by DFAIT at the outset of the process. The consultation also featured a significant on-line component. According to DFAIT's final report on the process, there were 600,000 visits to the Dialogue website, 28,000 downloads of the discussion paper, 2,000 registrants for the web forum, and "several thousand responses" submitted by mail, e-mail, or the web forum (DFAIT 2003, 3).

Another significant example of the Internet being used to mediate citizen consultation is the proceedings of the Romanow Commission on health care. Between April 2001 and November 2002 the commission engaged in a very ambitious and extensive program of public consultation. Along with the usual activities of research, public hearings, regional forums, expert workshops, round tables, and written submissions from individuals and groups, the commission made extensive provision for electronically mediated public input. These included a series of televised forums, the completion of over 13,000 on-line consultation workbooks, thousands of responses to nine different web-based surveys, the posting of official submissions and discussion papers on the commission's website, almost 2,000 calls to a toll-free telephone line, and over 4,400 individual correspondence submissions by e-mail (Commission on the Future of Health Care in Canada 2002, 259-300).

Of course, federal departments and agencies are also constantly engaging in various forms of consultation with experts, stakeholders, and partners, and many of these consultation processes are mediated at some point by digital network technologies. This mediation can take the form of electronic correspondence or information distribution, and sometimes the consultation itself can be mediated electronically via telephone, video conference, or some form of on-line communication. Thus, digital technologies certainly play a role in the routine consultative processes of government. The hard questions, however, are whether electronic mediation has been used significantly to engage the participation of a broader array of citizens in these routine processes, whether the form this participation has taken is substantially dialogic and deliberative, and whether the participation mediated in this matter has a consistent and discernible impact upon policy outcomes and decisions.

A number of federal government bodies have used digital technology in a variety of ways to enable public input on matters of public policy. Comprehensive, systematic analysis of these disaggregated endeavours has yet to emerge. Recently, the Department of Canadian Heritage, in conjunction with a number of other federal departments, launched the Consulting Canadians pilot project, a website designed to provide centralized access to government consultations currently under way. Visitors to the Consulting Canadians site are presented with links, indexed by subject area, department or agency, and title, to information detailing consultations under way in various parts of the government. The site lists only consultations submitted by the sponsoring agency or department. Users can also access a calendar of scheduled consultations and view summaries of results of past consultations. In most cases, users are directed to sites that set out the terms and means by which individuals or groups can make written submissions (either electronically or by post) to the sponsoring department or agency, or appear in person at scheduled hearings on the matter at issue. This suggests that while the Consulting Canadians site might serve to make more Canadians aware of a certain range of consultation exercises being undertaken by government, it does not enlist the properties of digital networks in a serious effort to make the consultation process *itself* more participatory (especially in a deliberative sense) or inclusive than has been the norm prior to the advent of the Internet.

The same might be said of the use of digital technologies by members of Parliament, who remain one of the most important points of contact between everyday citizens and the political operation of government. As Jonathan Malloy (2003) recounts in a recent study, digital technology has certainly affected the work routines of MPs, particularly with respect to their relationship with public servants. The Internet also, of course, mediates contact between citizens and elected representatives. Malloy's study documents a substantial increase in the volume of communication received by MPs from both constituents and "communities of interest" as a result of e-mail, and suggests that politicians and their staffs have yet to arrive at satisfactory routines for coping with this increased volume (18). A subsequent study sponsored by the Centre for Collaborative Government confirms that most MPs now make extensive use of ICTs, with a heavy emphasis on use of the Internet for e-mail, research, and gathering logistical information (Kernaghan, Riehle, and Lo 2003, 6).

Have MPs made significant attempts to utilize digital technologies to engage their constituents in policy development and decision-making processes that are more participatory, inclusive, and deliberative than those typical of previous media formations? The available evidence suggests they have not. In 2002 the Centre for Collaborative Government studied the web presence of all sitting MPs (Crossing Boundaries National Council 2004). The study found that 77 percent of MPs either had functional websites or had immediate plans to construct one. More interesting were the characteristic features of the MPs' sites. An overwhelming majority of MPs used their websites to disseminate basic information, such as personal biographies (94 percent), constituency newsletters (55 percent), contact information (93 percent), speeches (63 percent), and press releases (73 percent). A small minority, however, included more advanced tools on their websites that might enable enhanced citizen participation in various forms. Comparatively few MPs' websites provided for direct on-line feedback (37 percent), and even fewer made provision for on-line surveys or polling (8 percent), advanced search functions (10 percent), an electronic bulletin board (1 percent), and access to information such as how to submit petitions (7 percent) or on the MP's voting record (2 percent). According to the authors of this study, "The majority of MP websites are still using the internet as a top-down information tool ... MPs have done little to take advantage of potential interactivity."

The study by Kernaghan, Riehle, and Lo (2003, 9) confirms these findings. According to their survey of ICT use by a sample of 66 parliamentarians, 74 percent of MPs believe that the Internet is "important" or "highly important" for consultation with constituents. A closer look, however, reveals that MPs have generally declined to take advantage of on-line applications that might help make this consultation more interactive, two-way, and horizontal. Of those surveyed, only 23 percent had ever used the Internet to conduct an on-line survey, and only 15 percent had used the medium for an on-line opinion poll (10). These are relatively minimalist applications, easy to design and implement. When it comes to more demanding and substantive interactive applications, the numbers shrink dramatically. Only 9 percent reported ever having sponsored an electronic discussion forum, while electronic town halls and on-line chats had been tried by a mere 1.5 percent and 3 percent, respectively.

Widespread use of such applications may not result unambiguously in more participatory, inclusive, and responsive political representation. Still, the near-universal reluctance of MPs even to experiment with these possibilities signals what is probably a deeper reticence among elected representatives regarding the democratic potential of new media. This may be the reason MPs have thus far failed to leverage the distinctive properties of the Internet to the best democratic advantage of Canadians, opting instead to use the medium as they would any other broadcast technology, to "transmit information on a large scale to others without necessarily receiving information in return" (Kernaghan, Riehle, and Lo 2003, 6).

## Evaluating Government Use of ICTs

In her 2003 report to the House of Commons, Canada's auditor general evaluated the federal GOL initiative. Her assessment was less than glowing. While she praised the clarity of the government's vision for the project, and the horizontal process adopted to develop it, the auditor general criticized the government for having an inadequate plan for achieving this vision. In particular, she lamented the lack of a strategic plan detailing expected outcomes, identifying required resources, or providing a means for weighing the costs of the initiative (Office of the Auditor General 2003, 9). According to the report, despite the massive commitment of public resources to the project, "there is no clearly defined end state for GOL ... there has never been a comprehensive, well-articulated strategy or a consolidated strategic plan for the initiative" (10-11). This lack of specific, detailed expectations raises the prospect that "the government could declare victory in 2005 without having to measure its accomplishments against a set of clear expected outcomes" (11). Additionally, the GOL initiative has largely escaped the effective scrutiny of Parliament: according to the report, while Parliament has received adequate information on the GOL project vision and achievements, detailed information regarding costs, performance evaluation, and issues and risks has not been made available in a systematic fashion (17). The report goes on to detail a lack of planning with regard to the changes likely to result in departments and agencies most closely affected by the shift to electronic service delivery, and also criticizes the GOL initiative for failing to develop detailed plans for encouraging Canadians to actually make use of online services at a rate that would secure a reasonable return on the government's investment in this project (20-4, 28).

As significant as these criticisms of the GOL initiative are, they do not speak directly to the Canadian Democratic Audit's criteria of participation, responsiveness, and inclusiveness. The auditor general's report sheds welcome and careful light on the administrative, strategic, and financial integrity of the GOL initiative, but it does not address the broader question of whether the government's approach to the deployment of ICTs has contributed to a more responsive, inclusive, and participatory democracy in Canada.

One answer to this question stems from the fact that the defining preoccupation of the government's efforts in relation to the use of ICTs has been to enlist them as instruments in the rationalization of government service delivery, rather than as media of enhanced citizen participation in policy development and decision making (Longford 2002, 17). This paradigm casts citizens as clients, customers, and consumers seeking convenience and good value, rather than as participants in the political practice of public judgment; governance is reduced to the management of a dispensary. Here, notions of participation, responsiveness, and inclusiveness – at least in their distinctly *political* forms – pale under the brilliant glare of transactional efficiency, easy consumption, and cost-effective product delivery. Shopping malls, it could be argued, are relatively inclusive of the mass publics they invite to participate in choosing among the options arrayed for consumption, and vendors strive to be responsive to the appetites of their customers. If this is what we mean by democracy, then shopping malls are relatively democratic social spaces, and electronic service delivery is a highly democratic form of governance.

Charitably, it could be said that efficient on-line service delivery is precisely about democratizing access to government, albeit government understood in the truncated role of service provider. More critically, one might point out that ICTs and electronic service delivery have played a key role in a broader restructuring of the public sector, the effects of which have not been unambiguously positive for democratic public life in Canada (Crow and Longford 2000). The development of e-government in Canada has coincided with a comprehensive restructuring of the civil service to reflect the priorities of neoliberalism and the principles of "new public management": the size of the public sector (and its role vis-à-vis the market) has been dramatically reduced, and what remains is managed according to practices and norms borrowed from private enterprise. It is no accident that the Treasury Board's *Blueprint for renewing government services using information technology* and the federal government's program review – the latter involving widespread administrative rationalization, elimination or privatization of programs and services deemed nonessential, and extensive public sector job losses – were launched almost simultaneously in 1994 (Longford 2002, 9). As Graham Longford writes, when the state's role is "confined to that of effectively managing a more tightly circumscribed set of services, with a focus on efficiency, cost-effectiveness and 'value-added,'" the result is a "confused and impoverished conception of the citizen" (15; see also Brodie 1996).

E-government, understood as one-stop electronic service delivery and deployed as part of a broader strategy for rationalizing the Canadian state, does not contemplate citizenship as participation in policy development and decision making. Longford (2002, 20) captures this distinction perfectly: "While the ease and convenience of seamless service might be desirable for citizens wishing to *transact* with government for *existing* services, they may be less well served in situations where they wish to effect *change* to them by *influencing* government." Indeed, as he goes on to point out, the symbolic reduction of government to a series of "single windows" may provide efficient access to services, but may also undermine the civic literacy required for effective political citizenship, by obscuring the institutional complexity of government power and accountability.

Thus, even conceding the point that "ICTs are very likely to lead to more efficient service delivery, [it] is not at all clear that they will lead to a form of government that is more open, transparent, accountable or democratic than conventional government" (Lenihan 2002c, 8). This raises the question of the government's more explicit efforts with respect to e-democracy. Have these elevated the participatory, responsive, and inclusive character of democratic public life in

Canada? Here again the answer is ambiguous. The government's campaign to provide Canadians with on-line access to politically useful information cannot be casually dismissed as insignificant. Information is definitely an important participatory resource in a democracy and, to the extent the government has used the Internet to provide more Canadians with greater access to policy-relevant information, it is to be commended.

The government, however, has yet to arrive at a satisfactory appreciation of the status of information as a public resource (Crossing Boundaries Political Advisory Committee 2003, 9-11). As Longford writes,

> By saving us a trip to the library or government book store, on-line government certainly makes accessing information more convenient, for those with internet access, but it has not signaled a fundamental shift in government attitudes or practices with respect to disseminating information, particularly that of a politically sensitive or embarrassing nature. Indeed, in conflict with the stated goals of e-government, [the Canadian government] has embraced numerous policy and administrative initiatives in the 1990s which have actually *eroded* the informational rights of citizens (Longford 2002, 21, emphasis original).

Here, Longford refers to a range of measures, including the increased commercialization and commodification of information products and data-gathering services at agencies such as Statistics Canada; increased fees and decreased administrative support for access to information requests, in tandem with a more adversarial, obstructionist approach toward complying with them; and devolution (through outsourcing and privatization of service delivery) of information holdings to private contractors not subject to access to information laws (22). Together, measures like these contribute to an enclosure of the information commons that deprives citizens of necessary political resources and belies the government's own rhetoric concerning the spirit of openness, accessibility, and transparency supposedly animating its embrace of ICTs.

It may be true, as Donald Lenihan (2002c, 25) writes, that new technologies present "a unique opportunity ... to engineer a quantum leap in the quality and quantity of information that is available to support public debate and decision-making." But ICTs are no magic bullet. Absent a concrete commitment to provide material support for democratizing access in its *non*technological dimensions, rhetorical promises of a digital cornucopia of government information will continue to sound like ideological cover for the democratic liabilities of neoliberalism and the new public management.

In any case, the relationship between technologically mediated access to information and more, better-informed political participation is not a simple one. While good information is an indispensable resource for those already predisposed toward political participation, it is far from clear that more convenient access to greater volumes of information contributes to the activation of people who are otherwise disengaged politically. As I will discuss further below, very little evidence suggests that ICTs have served to mobilize the mass of politically inert citizens currently inhabiting Western liberal democracies (Norris 2001, 217-31), or that technologically enhanced access to growing volumes of information acts positively upon levels of formal political participation (Bimber 2001). This calls into question the proposition that the availability of information on-line has any independent effect upon political participation whatsoever.

There remains the question of whether the government's more direct attempts to use digital media to engage Canadians in policy development and decision making have contributed substantially to more inclusive, participatory, and responsive citizenship possibilities. Certainly, the democratic potential of these technologies should not be dismissed out of hand. As Lenihan (2002c, 27) writes, ICTs "could be used to extend public space in ways that might promote consultation and dialogue between citizens and their governments. Through this dialogue citizens and stakeholders might express their views, propose ideas, explore differences and participate more directly in decision-making ... It could contribute – perhaps very significantly – to the revitalization of democracy." But the government has not explored this potential with nearly the zeal it has brought to its deployment of ICTs in the rationalization of administration and service delivery. In fact, as the director general of e-services for Department of Canadian Heritage confided at the 2003 Information Highways Conference in Toronto, "Everybody will say engagement is important, but everybody is scared ... about doing it in government" (quoted in McQuillan 2003).

That being said, things may be changing. As outlined above, recent high-profile examples, such as the Dialogue on Foreign Policy and the proceedings of the Romanow Commission, suggest the government has become more interested in the possibilities of electronically mediated consultation. In 2003 Public Works and Government Services Canada established the Online Consultation Technologies Centre of Expertise, the purpose of which is to identify and develop effective tools and best practices for on-line public consultation in government. The assistant deputy minister of information technology services described the findings of the centre's inaugural report:

> More and more federal government departments and agencies are engaging in on-line consultation efforts as part of their public involvement activities. These consultation exercises include not only on-line feedback opportunities but increasingly on-line discussion-based opportunities for citizens' engagement. In surveys and interviews with government staff from a wide variety of departments, 56% of respondents reported conducting consultations using an on-line discussion based process, a statistic suggesting a movement toward a more interactive model of policy consultation with Canadians (Turner 2003).

This is suggestive, but the work of the centre is in its infancy, and there are little comprehensive data or systematic analysis concerning the extent and characteristics of electronically mediated government consultation in Canada. Even if there is a trend toward a *quantitative* increase in on-line consultation by federal departments and agencies, the crucial questions from the perspective of this audit concern the *quality* of the participation that will ensue, and the manner in which the government will respond to it.

These are empirical questions whose answers will require detailed study of actual proceedings, as they unfold. This evaluation will necessarily be informed by what we already know about the democratic value of existing government consultation practices, regardless of whether these are mediated by ICTs or not. As Susan Phillips and Michael Orsini have documented in a discussion paper commissioned by the Canadian Policy Research Networks, current government consultation practices in Canada leave a great deal to be desired from a democratic standpoint:

> Canada's primary institutions are not assuming as effective a part in citizen involvement as they might. In particular, political parties, members of Parliament and parliamentary committees play an unnecessarily weak role in involving citizens. The intergovernmental machinery, while increasingly important as a player, is as closed as ever to non-governmental actors. Although the public service has undertaken considerable public consultation, the consistency and manner in which this is done varies enormously across departments ... On major policy issues, the standard template of public consultation is deployed, complete with all the problems that have come to be associated with it – government controls the agenda and who is invited; information flows in one direction; and the process is episodic and ad hoc (Phillips and Orsini 2002, iii).

The democratic significance of technologically mediated consultation will depend ultimately upon the quality of government consultation efforts *in general.* If, as Phillips and Orsini conclude, effective means of citizen engagement will require fundamental reform, it will have to be accomplished notwithstanding the adoption of novel instruments of communication (iii). In short, the democratic prospects of on-line engagement are primarily a matter of ethos, and only secondarily a matter of technique.

Much hinges on the motivation behind a given agency's engagement and consultation strategy. At its most cynical extreme, the value of public consultation is derived from its mere appearance, which can be invoked to grant policy development and decision making undertaken by elites a veneer of democratic legitimacy they would otherwise lack. Engagement strategies can also be driven by the goal of "educating," or increasing the "awareness" of, the public and significant stakeholders with respect to a contentious policy issue, in the hope of mitigating opposition to impending outcomes in that area. Similarly, consultation can be an effective means of testing communication strategies surrounding particular initiatives, programs, and policies, and of gathering strategic information about how different presentations might be received by relevant constituencies. In each of these cases, consultation serves strategic, managerial, public relations – but not especially participatory – purposes (Chadwick and May 2003, 277). In its most democratic moments, consultation crosses the line into really inclusive, deliberative participation of citizens in policy development and decision making. For some, these are the qualities that distinguish genuine engagement from mere consultation (Mackinnon 2004). In any case, the point here is that ICTs can be enlisted in the service of any one of these distinct motivations.

Motivation also informs the actual process of consultation, procedurally and in terms of the conditions under which it takes place. To put it bluntly, participatory, inclusive, and responsive intentions are more likely to lead to participatory, inclusive, and responsive processes than are more managerial intentions. Consultation and engagement can take many forms. These range from the highly privatized, plebisicitarian registration of self-interest and opinion characteristic of referenda and polls, to somewhat more open forms of survey research and brief submission, to more deliberative forms that involve citizens and decision makers in direct, interactive dialogue with each other in the process of forming, rather than just asserting, collective preferences. Each of these has its place in the landscape of contemporary liberal democracy and, technically at least, ICTs are capable of serving any of them. Still, most of those who advocate an increased role for citizen engagement in governance have in mind these latter, more deliberative forms, and hope that ICT applications can be configured for these ends.

The characteristics of deliberative forms of on-line engagement are essentially the same as those of deliberation by other means: access to balanced information and expertise; a collectively determined agenda; adequate time for consideration of issues; freedom from manipulation and coercion; rules for discussion; inclusive participation; multidirectional discussion; and respect for differences (Coleman and Gøtze 2001, 6). The use of ICTs to mediate deliberative processes can sometimes present special challenges with respect to these criteria (for example, as I will discuss in Chapter 5, inclusivity can be undermined by differential levels of public access to network technology). ICTs can also introduce problems specific to the medium, such as the authentication or anonymity of participants, privacy, and records management. The criteria themselves do not change, however. And whether these criteria are satisfied depends less upon technical design than upon the willingness of consultation sponsors to embrace, rather than eschew, inclusive deliberation.

Ultimately, the democratic significance of public consultation by government, however it is mediated, depends upon the outcome of the engagement process in specific actions. Engagement that is not linked clearly and decisively to policy outcomes and decisions is at best a counterfeit of democracy, and at worst, it can undermine the legitimacy of those outcomes and decisions. Here, the issue is whether government agencies are, have been, and can be expected to be responsive to electronically mediated consultation with citizens. Again, this issue is not particularly technological. Instead, it comes down to the government's responsiveness to citizen participation in general.

A government that is unresponsive to its citizens off-line is no more likely to be responsive to their concerns when they are expressed via digital media. That being said, it is very difficult to discern and measure the degree to which policy and decision makers actually respond to the input they solicit in various consultation exercises, particularly those that are open and inclusive of a wide array of interests. Almost every document introducing a new government program or policy measure includes some reference to how citizen input helped to frame its general approach, and most try to bolster their particular decisions or recommendations with references to public submissions that supported that position. This was certainly true of the two prominent examples of widespread consultation cited earlier, the Foreign Policy Dialogue and the Romanow Commission, and also the CRTC hearings on convergence and new media discussed in Chapter 2. Such vague and strategic assurances, however, do not directly establish the actual impact of the breadth of citizen engagement on this or that final outcome.

This sort of responsiveness is very difficult to determine in the absence of access to, and systematic research upon, the inner workings of decision making at the upper ranks of closed government hierarchies. It would be a gross exaggeration to say that public input is completely inconsequential to the outcome of government consultations. The research that has begun to emerge in this area is not encouraging, however. Recent work on the Foreign Policy Dialogue found that the exercise provided only minimal opportunity for meaningful engagement *between* the public and official decision makers, despite the fact that, *among themselves*, citizens were "successful at developing, maintaining, and enforcing norms of civil discourse, and that these norms helped to promote understanding, tolerance and consensus-building" (Hurrell 2004, 1).

The government's own guidelines for consultation recognize and affirm some version of the democratic criteria being discussed here, including the need for clearly established connections between process and outcome, and so signal at least an appreciation of the value and demands of deliberative citizen engagement (Privy Council Office 2002). Of course, these are only guidelines, and there is considerable discretion and variation in their application. Most analysts agree that governments have a considerable way to go in establishing inclusive, deliberative citizen participation as a regular and prominent part of policy development and decision making, and that travelling this distance will require a concrete commitment to substantial, systemic change (Crossing Boundaries Political Advisory Committee 2003; Phillips and Orsini 2002; Chadwick and May 2003). New ICTs can conceivably assist governments in making good on such a commitment, but the commitment itself will have to be political rather than simply technological. And the fact remains that technological activity can also serve quite cunningly to obscure the absence of more concrete political reforms.

## Political Parties

Political parties are, in a manner of speaking, communication technologies. They are artificial instruments that mediate the flow of information between the party's leadership and its audience, the electorate. Information flows through parties in both directions: parties gather useful information about the electorate and communicate this to their leadership for strategic consideration; they also disseminate the messages of leadership to the electorate for its consideration, by a variety of means. Or, at least, they used to do these things. Already in 1979, John Meisel (1985) recognized that the communicative function of political parties had been more or less usurped by two other communication technologies. The role parties had once played in disseminating the political messages of their leadership (through, for example, local party meetings, events, and campaigns) had been replaced by television, through which messages, either as news filtered by journalists and editors or in the form of advertisements and televised events, could be broadcast directly to an audience of millions simultaneously from a central source. In this way, television "requires parties to centralize their informational activities" (Meisel and Mendelsohn 2001, 170). Similarly, the mediating role party members once played in gathering information and communicating it to leadership was overtaken by mass opinion polling, typically facilitated by electronic technology. This led Meisel (1985, 106) to conclude, even before the advent of the Internet, that "the party organization is no longer needed as an essential information network."

Changes in communication technology have long been associated with changes in the operation and orientation of Canadian political parties (Carty 1988, 15-30; Carty, Cross, and Young 2000, 178-210). Digital ICTs have become part of the dynamic identified by Meisel, whereby the communicative function of parties, or perhaps more specifically of party memberships and local organizations, has been made redundant. Parties still need members and local organizations for a variety of reasons, but not primarily for information gathering and dissemination, and not as media of political communication between the party and the broader electorate. These functions are increasing served by sophisticated communication technologies deployed by relatively centralized, strategic leadership within the parties. Television, of course, remains the most important technology of mass political communication in Canadian party politics, both during and between elections. The strategic value of ICTs has escalated in recent years, however, and they have become an important tool in the communicative practices of Canadian political parties.

The use of digital technology by contemporary political parties in Canada falls into three categories: internal administration and mobilization, publication, and data gathering and analysis. Like other large organizations, political parties require considerable internal communication for purposes of coordination and management. The fact that their membership is geographically dispersed, sporadically engaged, and seldom gathered in one location presents challenges for this sort of communication. The ability of ICTs to mediate asynchronous communication across great distances at great speeds and relatively low cost, delivering party information directly to members in their homes or workplaces, has made them very useful to party administrators. Membership databases, electronic mail and mailing lists, automated telephone dialling and messaging systems, and party websites are an efficient and cost-effective complement to traditional paper and postal correspondence (Alexander 2001a, 465; Kippen 2000, 11, 25-7). These facilities become especially important during election campaigns, when the need to coordinate strategy, manage logistics, distribute current information to party workers, solicit donations, and mobilize supporters to participate in events (and to vote) becomes paramount. Parties have also gradually begun to experiment with various combinations of television, telephone, and networked computer technology to enable their members to participate more directly in partisan events and processes, including electronic town hall meetings, telephone polls, policy discussion groups and, perhaps most significantly, the selection of party leaders (Cross 1998; Barney 1996; Courtney 2004, 121-2).

Parties have also come to recognize the utility of the World Wide Web as an instrument of mass, multimedia publication. By 1997 each of Canada's major federal parties had established a web presence, and each has steadily increased the content and sophistication of its site through the 2000 election to the present (DeRabbie 1998; Kippen 2000; Attallah and Burton 2001). Visitors to the website of a major Canadian party can expect to find some variation on the following:

- leader biography
- transcripts of leader's speeches
- press releases and news bulletins (current and archived)
- multimedia resources (images, graphics, audio, and video)
- roster of party MPs or candidates (often highlighting cabinet members or critics, with links to individual biographies or websites)

- party constitution
- list of party officers
- party policies, platform, and issue papers
- membership information
- on-line donation and volunteer forms
- contact information (including links to constituency associations)
- calendar of events and leader's itinerary
- subscription to electronic mailing list or newsletter
- on-line feedback forms
- on-line polls or surveys
- links to provincial party organizations
- links to basic government information.

Websites of individual candidates or elected representatives contain similar information, with an emphasis on local issues, events, and resources. Like all other elements of party machinery, the importance of party websites is elevated during (and just before) election campaigns. During these periods, membership recruitment, fundraising, and volunteer activities experience a spike and, more importantly, elections see the party engaged very actively in the dissemination of information to its activists, the press, and voters.

This dissemination occurs via several means, but official websites offer parties considerable utility for campaign communication. At a fraction of the cost of a comparable amount of paid television, radio, and print advertising, parties can deliver extensive, detailed campaign information to journalists and voters on the web twenty-four hours a day, reacting quickly to issues, crises, and opportunities as they develop (Kippen 2000, 20-3). Using their websites, parties can communicate directly with voters without the filtering, framing, and interpretation that is the price of delivery through mass media outlets (Alexander 2001a, 470). Centralizing information dissemination through the party website (or even via electronic mailing lists) also allows campaign strategists and communication directors to exercise a high degree of control over the content and shape of campaign messages. Canadian parties have also begun to experiment with on-line advertising, buying space for click-through banner ads on mainstream media and political sites and, in some cases, employing the novel technique of "buying" keywords on popular Internet search engine sites that, when entered into the engine's search field, trigger the display of an ad for the party alongside the results of the search (Attallah and Burton 2001, 228-9; Alexander 2001a, 463).

Still, parties remain unsure as to whether sophisticated on-line information dissemination techniques and strategies are worth the investment. The fact remains that party websites and on-line ads reach a relatively small audience of politically engaged professionals and citizens, and the impact of their content on voter preferences is unclear. Perhaps for this reason, television – news coverage, leadership debates, and advertising – remains the most significant medium for the mass dissemination of partisan information during election campaigns (Attallah and Burton 2001, 215).

Networked, computerized ICTs have, however, been unambiguously beneficial to parties as instruments of data gathering and processing. Electronic databases allow for the storage, search, and retrieval of massive volumes of complex information, and increasingly powerful computers enable increasingly sophisticated processing and modelling of this information. When asked about important partisan uses of digital technology in a recent survey, federal parliamentarians who identified database management (72.7 percent) and voter targeting (64.6 percent) outnumbered those who identified on-line campaigning (48.5 percent) and on-line fundraising (28.8 percent) by a considerable margin (Kernaghan, Riehle, and Lo 2003, 11). Similar views were expressed at a special session on political parties and new technologies at the 2003 Crossing Boundaries conference in Ottawa (Crossing Boundaries National Council 2003). Here, strategists from the New Democratic, Liberal, and Canadian Alliance parties spent considerable time highlighting the utility of ICTs for profiling and tracking voter preferences; they were less animated by the possibility that these technologies might provide new means for party members and voters to participate more deliberately and dialogically in party politics.

And so it comes as no surprise that parties have seized on these technologies in an effort to craft campaign appeals aimed at the political tastes of particular regions, demographics, and even individual voters (Carty, Cross, and Young 2000, 208-9; DeRabbie 1998). Combining data from Elections Canada's permanent electronic voters list – established in 1993 and described as "the single most important technological innovation in Canadian politics this century" (Kippen 2000, 10) – with constituency information gathered by the party and databases purchased from commercial sources, strategists are able to generate a multiplicity of specific appeals targeted at narrow categories of identifiable groups and individuals, whose responses can subsequently be tracked and incorporated into further campaign refinements. In short, new technologies and techniques of database management and data mining have made it possible for parties to "customize, customize, customize" their campaigns (Alexander 2001a, 467).

## Evaluating Parties' Use of ICTs

Has the use of ICTs by Canadian political parties made them, or the democracy they serve, more responsive, inclusive, and participatory? This question can be answered only in light of an assessment of the democratic character of recent party activities more generally. As Carty, Cross, and Young have documented, the populist challenge posed by the rise of the Reform/Alliance party stimulated contemporary Canadian political parties to gesture in the direction of democratization (Carty, Cross, and Young 2000, 107-29). These gestures have taken forms ranging from experiments with electronically mediated participation in party policy forums to direct election of leaders by party members, but their democratic significance is debatable. For example, the Reform/Alliance party's forays into "teledemocracy" may actually undermine deliberative, public-spirited democracy by using plebiscitary instruments to minimize the effectiveness of organized interests, by appealing directly to individuals and their privately registered opinions (Barney 1996; Barney and Laycock 1999). And the move from delegated to direct election of party leaders has not been an unambiguous democratic gain, as the expenditure of power, resources and influence once applied to delegate selection contests has now been transferred to the recruitment of masses of new members whose sole involvement in the party is the minimalist act of voting for the machine that recruited them. As Leonard Preyra has observed, "Parties have created new leadership selection processes that *appear* to make leadership selection more inclusive, empowering, transparent and accountable. However, at the operational level a huge chasm in expectations and outcomes remains" (Preyra 2001, 455, emphasis added).

Given parties' relatively halting approach toward democratization in general, it is not surprising that their specific efforts with ICTs have been less than transformative. The use of electronic communication media such as e-mail to coordinate administration and to facilitate mobilization, recruitment, and fundraising have undoubtedly made parties more efficient in terms of internal communication. Similarly, parties' use of the web to publicize party and campaign material probably means that more people, both inside and outside the party, have better access to unmediated partisan information than was previously the case. Administrative efficiency and effective publicity, however, do not necessarily make for more inclusive, participatory, and responsive democratic processes. Indeed, parties have been very reluctant to pursue with vigour and creativity the potential that ICTs present for the mediation of more routine, deliberative, participatory exercises explicitly connected to party policy, either among their own memberships or more widely.

Where parties have been quick to seize upon the potential of these technologies is in their usefulness for sophisticated gathering, storage, and processing of data about voters and their preferences, in an effort to craft campaign strategies that are at once highly centralized and highly customized. Some might argue that such practices enable parties to be more precisely responsive to the particular preferences of citizens. This may be true, but only if we accept that the techniques of customer-relations management – rather than the norms of inclusive, deliberative dialogue among party members, voters, and party elites – provide an adequate framework for political communication between parties and citizens. Perhaps it has been a long time since parties have systematically practised anything approaching the standard implied by these norms, but this fact should not lead us to dignify data mining, or the use of the web to provide access to interactive electronic brochures, with a name it doesn't really deserve.

## Advocacy Groups and New Social Movements

A discussion of the political uses of ICTs that focuses on governments and traditional actors such as political parties may arguably miss the point entirely. New ICTs, it has been suggested, mediate an entirely new brand of politics in which nonstate interest groups and new social movements – loosely affiliated, noninstitutionalized, geographically dispersed, dynamic coalitions of individuals and organizations gathered around particular issues – play a central role (Castells 2001, 138-55; Webster 2001; Castells 1997). New social movements in civil society and new information technologies are thought to be particularly complementary. First, the new politics is said to be a highly "informational" or symbolic brand of politics enacted in and through the space of technologically mediated mass communication. New media such as the Internet provide activist groups and coalitions with a level of access to the mediated public sphere that they were often denied in relation to traditional mass print and broadcasting media. Second, there is symmetry between the decentralized, nonhierarchical, networked architecture of digital communication technologies and the organizational structure and relationships of new social movements, which are themselves typically organized as networks (Castells 1997; Young and Everitt 2004).

Considerable opportunities are afforded to interest groups and new social movements by digital networks. Applications such as e-mail and multimedia, hypertextual websites provide civil society actors with the means to accomplish a range of functions that are crucial to their operations and impact:

- collection, production, publication, and archiving of information resources
- event promotion, recruitment, fundraising, and the solicitation of other forms of support
- consciousness raising, political education, and membership training
- networking with sympathetic and allied organizations and coalitions (including transnational activist networks)
- internal organization, administration, mobilization (e.g., "action alerts"), and coordination of activities
- mediation of democratic dialogue, debate, and deliberation
- access to independent media sources, news reporting, and alternative journalism
- participation in state-sponsored consultation processes
- development of new repertoires of political action (e.g., "hacktivism," mass e-mail campaigns, denial of service attacks, electronic petitions, website defacement, parody sites, etc.) (Barney 2004, 126).

A growing case-study literature documents how advocacy groups and new social movements around the world have used digital networks in some or all of the ways listed above (McCaughey and Ayers 2003; Pendakur and Harris 2002; Webster 2001). These studies also reveal the liabilities of high-technology organization and activism. Some of these include the technological, financial, and time and labour resource burdens of mounting and sustaining effective digital operations, issues of privacy and surveillance, overreliance on a technology to which access is far from equally distributed, vulnerability to technology failure, and issues of censorship or freedom of expression and organization credibility. In general, however, most observers feel that ICTs have done more to help than to harm the operations and effectiveness of advocacy groups and new social movements.

The affinity between ICTs and the politics associated with new social movements is often bound up tightly with the transnationalization of political issues, authority, and activism, and the growing importance of global civil society networks and global public spheres. The supporting role that ICTs have played in the establishment and maintenance of the transnational antiglobalization movement – first dramatized in 1999 in coordinated global opposition to the Multilateral Agreement on Investment and reinforced in subsequent protest actions and the proliferation of independent media outlets – is typically held up as definitive evidence of the democratic political character of these technologies (Deibert 2002a; Potter 2002; Smith and Smythe 2001). ICTs have also assumed a central position in the activities of advocacy groups and social movements that are localized at the national or subnational level in Canada. Groups such as the Council of Canadians have played central roles in organizing national and transnational ICT-mediated activism, and it would be impossible to estimate the number of politically active organizations and social movement actors in Canada who make regular, creative, effective use of the Internet for some or all or the purposes outlined above. One site listing links to progressive Canadian advocacy groups and movements active on or via the Internet lists hundreds of organizations, and this is just the tip of a much larger iceberg (Jay's Leftist and Progressive Internet Resources Directory 2004). In all cases, use of ICTs is keyed explicitly to political action, whether this action takes place on-line or off-line, around single issues or broadly based interests, in reference to formal state authority or regardless of it, independently or in concert with national or global efforts. In short, ICTs have been an indispensable instrument in the political activities of contemporary social movements and advocacy groups.

## Evaluating ICT Use By Advocacy Groups and Social Movements

Has the embrace of ICTs by advocacy groups and new social movements made them, and Canadian democracy more generally, more inclusive, participatory, and responsive? In the above discussion of the use of ICTs by parties, I suggested that administrative efficiency and public relations should not be mistaken for any of these three criteria. But this warning should be modified with respect to advocacy groups and social movements. The reason is that among those who have benefited most from ICTs are advocacy groups and actors that have been, and continue to be, either formally external to the institutional operation of power in Canada, or otherwise politically marginalized. One of the dominant themes in the literature surrounding the political use of ICTs in Canada is the relative independence, in terms of communication, they afford silenced, disempowered, oppositional, and minority constituencies. A list of such constituencies might include indigenous peoples and their organizations, the women's movement, the peace movement, environmentalists, diasporic and ethnic minority communities, youth organizations, gay and lesbian organizations, community activists, and a wide range of groups, organizations, and movements engaged in political action around particular issues of common interest (such as health care, freedom of expression, economic justice, education, and racism).

Of course, groups with ties to government, or which are situated in the social and political mainstream, also make use of these technologies. But one of the peculiarities of the Internet is the bandwidth it provides to those voices and interests that have historically been excluded from mainstream media, discourse, and political influence. New media technologies offer these marginalized, activist groups a means of more or less autonomous organizational and mass communication that they have been denied in a communication landscape dominated by broadcast and mass press technologies to which access (on the production end) is comparatively limited. To the extent that digital networks have reduced the costs, and increased the effectiveness, of political organization, mobilization, publication, and networking for oppositional and marginalized social groups, they have certainly contributed to a somewhat more inclusive political space in Canada.

The benefit of increased pluralism, however, should be weighed against the possibility of increased fragmentation and balkanization of the public sphere. Just as the technicalities of digital networks enable more diverse interests to enter the space of public communication, so too do these technicalities make it possible for people to customize their political encounters – at least those involving information and communication activities – in ways that diminish their exposure to issues, interests, and opinions that diverge from their own, or from those of the groups to which they belong (Barney 2003, 112; Sunstein 2001, 51-79; Bimber 1998). The accelerated pluralism enabled by ICTs can make for a public sphere that is more inclusive of people with diverse backgrounds, affiliations, and interests, but perhaps also one in which people's actual encounter with diversity is not significantly enhanced.

Advocacy groups and social movements have also been more diligent than governments and political parties in their efforts to use ICTs for explicitly participatory purposes. Political participation that is more engaging and meaningful than that provided by traditional institutions is central to the identity of advocacy groups and social movements, particularly those occupying positions outside the political mainstream. Consequently, civil society organizations and actors have consistently oriented their use of ICTs toward enhancing the participation of their members both quantitatively and qualitatively. Network technologies are used to organize, mobilize, and support participatory action off-line, and also to mediate significant participatory practices on-line. These include the new forms of political "hacktivism" that the Internet has added to the repertoire of activist organizations, as well as the consistent efforts by such groups to deploy new communication technologies – using a variety of website and e-mail applications – to facilitate political dialogue, debate, deliberation, and decision making among their members. Additionally, independent media organizations such as Indymedia have used ICTs quite effectively in their efforts to establish a communication commons that provides a critical alternative to corporate-dominated mass media and news coverage (Kidd 2003).

Advocacy groups and social movements, at least those on the ideological and material margins of the liberal-capitalist mainstream, have responded somewhat more courageously than traditional actors to the opportunities and risks presented by ICTs, using them where they can to enhance and enlarge participatory opportunities. This courage may be linked to the habits of democratic responsiveness developed by interest groups and social movements as the basis of their legitimacy and vitality.

## Individual Citizens

On one level, the question of whether the use of ICTs has contributed to a more inclusive, responsive, and participatory democracy in Canada is answered by the manner in which most Canadians use these technologies in their everyday life. To answer this question in the affirmative, there would have to be some indication that these technologies serve as a medium whereby a significant number of Canadians engage in activities that can be categorized as distinctly political. For the minority of Canadians who are politically engaged – whether as members of the public service, political parties, interest groups and social movements, or simply as independent citizens – the Internet has become an indispensable medium, a fixture of their regular, lived political practices. But most Canadians, most of time, do not use the medium for distinctly political purposes. Any conclusions about the political significance of ICTs must be considered in light of this fact.

According to Statistics Canada's 2003 Household Internet Use Survey, 62 percent of Canadian households have at least one member who uses the Internet regularly (Statistics Canada 2003). What do Canadians do on-line? The survey lists e-mail as the most popular online activity in Canada, with nearly 49 percent of regular Internet users reporting regular use of this application. This is followed by general browsing (46.1 percent), searches for medical or health information (32.8 percent), travel information and service (30.4 percent), government information (29.2 percent), news (27.2 percent), electronic banking (26.2 percent), playing games (25.7 percent), education and training (24.3 percent), obtaining music (24.3 percent), and finding sports (23.8 percent) and financial information (23.5 percent). And while a few of these categories might be interpreted as signifying political uses, they are not necessarily so: the vast majority of e-mail is personal and work-related; those visiting government websites do so most often for reasons of education, research, and service retrieval; and while news gathering is a core activity of most citizens, its political value is difficult to gauge, especially when "news" in the contemporary climate means celebrity gossip and crime or disaster reporting as often as it does analysis of the latest federal budget. In any case, this survey supplies little evidence to suggest that distinctly or robustly political uses of the Internet rank highly for many Canadians.

Recent figures from other sources tell a similar story. Statistics Canada's 2000 General Social Survey, for example, lists the most popular subjects for Canadians using the Internet to search for information (in Crowley 2002, 475). The results are as follows: arts/entertainment/sports (56 percent), travel (45 percent), business/economic news (34 percent), work/job searches (30 percent), telephone listings (27 percent), computer information (27 percent), community services and events (26 percent), and government labour market information (10 percent). A recent study of youth Internet use conducted by the Media Awareness Network and Environics Research shows that they (and perhaps the future) are no more politicized in their habits with this medium than are the present generation of Canadian adults (in Crowley 2002, 478). According to the study, the main Internet activities reported by Canadian youth are as follows: downloading music (57 percent), e-mail (56 percent), surfing for fun (50 percent), playing games (48 percent), getting non-homework-related information (41 percent), instant messaging (40 percent), chat rooms (39 percent), and homework (38 percent). It is possible, of course, that Canadian youth use e-mail to organize politically, or that they search for political information when they are not doing their homework, but it is unlikely that they do so in significant numbers. Again, there is not much evidence here that a substantial percentage of Canadians, young or old, are using the Internet for the purpose of meaningful political participation.

Studies like these may simply fail to capture the political activity that Canadians do engage in using ICTs. Perhaps if these use surveys included specific items such as "to engage in political discussions" or "to contact my member of Parliament" among possible uses, they would show higher levels of political activity. Recent research into Internet use surrounding elections sheds some light on this possibility. Despite historically low levels of voter turnout in recent elections, general elections remain catalysts of increased levels of awareness and mobilization among ordinary Canadians. If significant levels of Internet-mediated political activity were to occur, it is reasonable to expect that it would be in the period surrounding elections, and that it would be evident to those who looked for it specifically. The Canadian Election Study team asked several direct questions about Internet use during the 2000 federal election campaign. When asked whether they used the Internet to get any information about the federal election, only 10.4 percent of those surveyed answered "yes" (Canadian Election Study 2000). A mere 4.8 percent (just under half of those who had used the Internet at all) reported having visited a party website during the campaign. Even fewer – just 1.5 percent – had e-mailed a message to a candidate or party during the campaign. In other words, the number of Canadians who used the Internet to participate even minimally in politics, during the period when they are supposedly at their most politically engaged, was very, very small. And if the number of Canadians who use ICTs in these minimalist, mainstream ways during periods of heightened political awareness and activity is this small, then there is little reason to doubt the statistics cited above: when it comes to how most Canadians use the Internet most of the time, politics barely registers.

These findings seem to cast doubt on claims that digital networks are dramatically rejuvenating democratic participation at the individual level, or that they serve to include a broader range of individuals in the practices of citizenship than was previously the case. In this, Canada is not alone. Studies in the United States and elsewhere confirm that while digital technologies have served to reinforce existing patterns of political mobilization and engagement, even amplifying the voices of the small minority of citizens who are politically active, they have done very little to mobilize the massive numbers of depoliticized individuals that increasingly populate Western liberal democracies (Norris 2002; Bimber 2001; Norris 2001, 217-31). Those who are already politically engaged or active have definitely integrated ICTs into their citizenship practices. Nevertheless, easier access to political information and communication is insufficient to stimulate significantly increased levels of participation, even in its most undemanding forms. In fact, those who characteristically participate vigorously in political life appear to do so despite, not because of, new ICTs. The medium, at least when it comes to the conditions that support individual political participation and engagement, might not be the message after all.

## When the Medium Isn't the Message

The foregoing review of ICT use by governments, political parties, advocacy groups, social movements, and individuals suggests that the mere presence or adoption of these technologies is not enough to guarantee a more inclusive, participatory, and responsive democracy. To be sure, these technologies can be deployed and used in ways that contribute to this outcome. But the possibility of this outcome relies less on properties inherent in the technologies themselves than it does on the political will of those using them.

For the most part, strengthening democracy has not been the primary goal animating the use of these technologies by political institutions and actors in Canada. Government, for example, has concentrated its efforts with ICTs on improving the efficiency and cost-effectiveness of service delivery, and on optimizing the administrative potential of e-government. It has paid comparatively little attention to the potential of new technologies to mediate novel forms of serious citizen participation in policy development and decision making, limiting its efforts in this area to minimalist, tentative gestures. Similarly, political parties have been quick to seize on ICTs as instruments of public relations and campaign strategy, but less than enthusiastic in experimenting with them as a means to make parties the site of more robust democratic participation. Perhaps this is because they have some insight into the nasty little fact that the overwhelming majority of individual Canadians in general do not use the Internet with politics in mind.

The sole exception in this regard is advocacy groups and social movements, which have led the way in terms of creative and progressive democratic appropriation of digital ICTs. This is not because they know something that the other actors and institutions discussed here do not know. Instead, it is because the normative commitment to democracy that defines most interest groups and social movements precedes and frames their approach to the use of ICTs. Absent such a prior and deeply held commitment, ICT use is unlikely to lead in unambiguously participatory, inclusive, and responsive directions. At the end of the day, no technological magic will be able to conjure an appetite for democracy in those who don't already have the stomach for it.

## Chapter 4

### Strengths

- Significant, creative use of ICTs has been made by advocacy groups and social movements, especially marginalized ones.
- New ICTs provide important space for alternative media and information providers.
- Increased critical attention is being paid to the potential of electronically mediated democratic engagement in Canada.

### Weaknesses

- Efficient service delivery eclipses democratic engagement as a goal in the Government On-Line agenda.
- Canadian political parties have not used ICTs to their full potential as means of citizen engagement.
- Most Canadian citizens do not use ICTs for political purposes.

# 5 DIGITAL DIVIDES

The word "democracy" is a name for a particular distribution of power. Specifically, it names a situation in which political power – the power to judge, to decide, and to act with public authority – is distributed equally among citizens. Democracy is a difficult standard that few, if any, contemporary political systems have managed to reach. But this difficulty does not necessarily invalidate democracy as a normative ideal toward which we might aspire, or as a critical standard against which we might measure current political practice. For, while it doesn't make much sense to use this standard to categorize political regimes as either democratic or not, they can be usefully categorized as *more* or *less* democratic in their institutions and functions. We can say that a political order is more democratic when it tends toward an equal distribution of power, and less democratic when it tends away from it. In this chapter, I wish to explore the relationship between new information and communication technologies and the distribution of power in Canada. Here, again, this relationship should be considered in both its possible directions: ICTs potentially affect the distribution of power in Canada, and the distribution of power in Canada also affects the democratic potential of ICTs. And so we must ask both whether ICTs have contributed to an equalization of power, and also how the existing distribution of power in Canadian society has affected their potential to make this contribution.

Specifying equality, particularly equality in relation to the distribution of political power, as the core of democracy casts the criteria employed in the Canadian Democratic Audit in a particular light. Participation, inclusiveness, and responsiveness are, without question, crucial attributes of any democratic political order, and no regime that fails to exhibit these can be called a democracy in anything but the weakest sense. What makes these criteria specifically *democratic*, however, is their tendency toward political equality. They can pertain under conditions of concrete inequality but, in these instances, they are easily deployed as ideological cover for the absence of substantive democracy. Participation under conditions of radical inequality, for example, can be used as "evidence" that a regime is democratic, when in fact it is not. When these attributes occur under conditions of equality, or when they make a material contribution to its realization, then participation, inclusiveness, and responsiveness assume their democratic aspect.

Thus, the question of ICTs and the participatory, responsive, and inclusive character of Canadian democracy cannot be divorced from the question of ICTs and political and economic equality more generally. A comprehensive examination of this question would require a volume of its own. Here, my hope is to introduce some issues that were not necessarily raised by the discussion of ICTs as either an object or instrument of democratic politics in the preceding chapters. These issues speak to the broader question of the role technology plays in establishing the material context that either supports or undermines democratic politics and citizenship. In this light, we will consider issues surrounding the digital divide, the political economy of digital technology, and ICTs and the democratic public sphere.

## The Digital Divide

Tracking the progress of access to the Internet, and studying the characteristics associated with access, has become a staple of sociological analysis of the impact of ICTs. In societies such as Canada's, where ICTs have become increasingly important for participation in mainstream social, economic, and political life, differential levels of access to these technologies can be a significant source of inequality, forming a "digital divide" between those who have access and those who do not. Colloquially, this inequality is often described as a separation between information haves and information have-nots, but this expression does not do justice to the complex relationship between these technologies and inequality. First, differential access to ICTs *reflects* existing socioeconomic inequalities as much as constituting them, and second, connectivity is only the most obvious way in which power surrounding these technologies is unequally distributed.

By 2003 roughly 68 percent of Canadians reported having access to and using the Internet, mostly at home, at work, or at school (CRTC 2003, 115). If the digital divide is understood primarily in terms of connectivity – i.e., the combination of access and use – then identifying the characteristics of those who are "unconnected" is an important democratic problem. In a Statistics Canada report published in 2001, the diagnosis was as follows:

> Internet users differ from non-users in average age, education, and income. Non-users of the Internet are more likely to be older individuals, and are more likely to have less education and lower household income than Internet users. Non-users are more likely to be women than men at every age group. Francophones are less likely to use the Internet than Anglophones, and those living in rural Canada are less likely to use the Internet than urban dwellers. When non-users were asked to identify the greatest barrier that keeps them from using the Internet, cost was cited by the largest percentage of the people. Lack of access to computers or the Internet was the second most often cited barrier (Dryburgh 2001, 4).

In other words, differential levels of Internet access and use generally mirror, and so presumably reproduce, existing indexes of socioeconomic advantage and disadvantage in Canada. Study after study confirms this basic pattern: "We see differences based on income, education, geography, gender, age, disability and aboriginal status. What is disconcerting ... is the suggestion that these gaps in Canadian society are widening. What is more, many of these divisions overlap, so that some groups are doubly or triply disadvantaged" (Looker and Thiessen 2003, 2; see also Rideout 2000). Of course, the specifics of some of these measures have changed over time; for example, lack of interest has now eclipsed cost as the main reason cited for not having Internet access at home (Reddick and Boucher 2002, 12-14). The basic finding, however, remains unchanged: "There has been marginal growth in internet access for all groups, but the digital divide has not narrowed" (i).

This picture of systematic inequality becomes more complex when one considers that mere connectivity only superficially captures the relationship between ICTs and social, economic, or political empowerment. While a lack of access to ICTs is certainly a source and marker of significant inequality, it does not necessarily follow that simply being connected constitutes empowerment in relation to these technologies. As several commentators on the digital divide in Canada (and elsewhere) have pointed out, in taking the measure of this problem we must consider dynamics that are more complex than simple technical access to, or periodic use of, an Internet connection (Paré 2004; Reddick and Boucher 2002; Clement and Shade 2000; Shade 1999). Put bluntly, not everyone who has access to this medium uses it as an equal. Even after connectivity is established, people experience serious inequalities in terms of *how* they use this medium (as passive consumers or as active contributors); in their *capacity* (skills and literacies) to use these technologies in ways that contribute to, rather than diminish, their autonomy; and in their *ability to influence* (either individually or collectively) the development, design, content, and regulation of the medium and its applications.

In other words, while for some people access to the Internet is a source of empowerment, autonomy, and agency, for many it simply means connection to a technological infrastructure in relation to which they remain significantly disadvantaged and powerless, *despite* their technical access to it. It should come as no surprise that those who suffer on the down side of these other digital divides also tend to occupy perennial categories of systemic disadvantage and inequality in Canadian society. Thus, for example, while most studies show that the differences between men and women in terms of connectivity have largely been erased in Canada, there remain serious discrepancies between men and women in terms of characteristic use, capacity, and power relative to the medium itself (Shade 2002, 73-92; Balka and Smith 2000).

A recent comprehensive survey of ICT use among Canadian youth sheds light on this situation. It showed that young men were more likely than young women to have written their own computer programs, that men spent more time programming than women, and that men were more likely than women to have played computer games, entered and analyzed data, or used graphics programs, spreadsheets, and CD-ROMs (Looker and Thiessen 2003, 8). The only activity in which young women's use outpaced that of young men was e-mail. As the authors of this study put it, "In other words while males and females report similar *levels* of use, males tend to use computers and ICTs in more *diverse ways*. Further, for any given type of task, males are more likely than females to say they use ICTs for this task every day. These diverse skills are ones that would serve the young men well when applying for high-skilled jobs using ICTs" (9, emphasis added). Young men also scored higher than young women on several attitudinal dimensions in relation to computers, with a significantly higher proportion of men than women agreeing that computers were important, fun, or interesting. Tellingly, while 38 percent of young men assessed their own skills with computers as "excellent," only 17 percent of young women offered the same self-appraisal (8).

Thus, while levels of the access to the Internet have equalized between men and women, it seems that computers and ICTs are still more or less "a guy thing." And so we might expect that the underrepresentation of women in positions of power relative to these technologies, and overrepresentation in situations where their encounter with these technologies is less self-determined and negotiable, will continue. Equal access to the Internet has not changed the fact that women are more likely to be found in call centres and teleworking (often after a full day of unpaid domestic labour) than they are to be in engineering laboratories, computer science classrooms, corporate boardrooms, positions of regulatory authority, and political office (Gurstein 2001; Stewart Millar 1998; Spender 1996). It has also not changed the fact that, due to the ubiquity of pornography, the persistence of sexual harassment, and the dominance of distinctly male norms of discourse on-line, "the internet and World Wide Web are actively and aggressively hostile to women" (Eubanks 2000, 1).

This is not to say women do not use ICTs in ways that express and enhance their autonomy. Those women who enjoy socioeconomic situations that enable them to control the terms and context of their use of these technologies have certainly used them to great advantage. And, like other advocacy groups and social movements, women's and feminist groups have made great use of ICTs to enhance their agency through publication, political organization, and mobilization, and the construction of venues for communication that accommodate and liberate women from the insecurity and discrimination to which they are subjected in other contexts (Gerlach and Hamilton 2002; Shade 2002, 33-53, 93-106; Sassen 2002). These activities are not insignificant, but little evidence suggests they characterize the manner in which most women experience ICTs most of the time. The point here is that even when the problem of unequal connectivity and access is solved, and despite the empowering use of ICTs by some women, the problem of a deeply gendered digital divide remains. This suggests, among other things, that state efforts to overcome the digital divide that focus solely on connectivity are likely to fail to address the deeper causes and manifestations of inequality in relation to ICTs and their use (Dowding 2002).

Similar analysis could be undertaken with regard to the encounter between Canada's Aboriginal people and ICTs. In terms of connectivity, it is generally conceded that Aboriginal people in Canada suffer from unequal access to the Internet, especially high-speed broadband services, as a result of multiple disadvantages: the geographic remoteness of many Aboriginal communities, relatively high incidence of poverty, and generally lower levels of formal education. All of these are significant variables in relation to Internet access and use. That being said, informative statistics regarding Aboriginal Internet access and use are difficult to come by. The Aboriginal Peoples Survey, conducted as part of the 2001 census, asked people in select Canadian Aboriginal communities whether they had used a computer or the Internet in the past twelve months (Statistics Canada 2004). These questions do not really tell us much, as a single use of the Internet at a community access terminal would register as a "yes" even though it does not represent significant connectivity (i.e., regular use at home or school). Furthermore, data were gathered only from communities that consented to being surveyed, and are not aggregated at the provincial or national level, making reliable generalizations quite difficult.

In 2002 the Department of Indian and Northern Affairs published a report on the level of connectivity in Aboriginal communities in Canada, but the data produced here are inconclusive at best. For example, 91 percent of Aboriginal communities are found to be connected to the Internet "in some way," though this includes community access via a single machine in a school, administrative office, or community centre, sometimes with long-distance charges. The report also finds that 89 percent of Aboriginal communities "have the telephone infrastructure necessary to connect to the internet at the household level," but this establishes only that a community has telephone lines going to individual households, not whether they are active or used for connection to Internet services. The study provides little real insight into rates of access or use in Aboriginal communities, and does not report on Aboriginal people living in non-Aboriginal communities and major metropolitan areas.

Nevertheless, the consensus appears to be that Aboriginal people and their communities suffer at the wrong side of the digital divide in terms of connectivity. This problem has motivated literally hundreds of federal and provincial projects and initiatives aimed at overcoming the Aboriginal connectivity gap (Department of Indian and Northern Affairs 2002, apps. A and B; Alexander 2001b). Aboriginal activists and organizations have recognized that more than a purely technical effort is required. As stressed in the final report of the 2003 Connecting Aboriginal Canadians Forum, closing the digital divide for Canada's Aboriginal people will require not just infrastructure but coordinated efforts at skills development and capacity building, support for the generation of digital content for and by Aboriginal people, support for appropriate economic and social development surrounding these technologies, and recognition of the need to engage Aboriginal communities and citizens themselves in the development of this technology (Connecting Aboriginal Canadians Forum 2003).

Only in the context of such a broader effort to provide and sustain conditions in which Aboriginal people might determine the manner in which they and their communities encounter ICTs will closing the connectivity divide become meaningful. Otherwise, connectivity will simply mean wiring Aboriginal people and communities into a technical system that is likely to perpetuate their disadvantage and reduce their cultural security. As James May (1998, 222) has written, "Television, computing networks, publication, telephones, and telecommunications are all taking a toll on traditional indigenous cultures." Mass communication technologies deployed under the control of Western capitalist and state elites have always exacted an assimilationist price upon Aboriginal languages and cultural autonomy, and the Internet is no exception (Nolen 2000). Gail Valaskakis (2002, 403), in her work on Northern communication media and Canada's Inuit, documents a history in which "non-Natives gained over-riding authority through mere possession of technology and control over its distribution." She describes a dynamic of "cultural replacement" in which economic and political dependency was supported via communication media that marginalized traditional modes of communication and indigenous content (405). This suggests that, when it comes to the manner in which communication technologies are implicated in the distribution of power vis-à-vis Canada's Aboriginal peoples, the "digital" part may be fairly new but the "divide" part is not.

Valaskakis (2002, 413) also documents how the Inuit have struggled historically to appropriate modern communication media for their own purposes and benefit, however, and suggests this struggle will not abate with respect to the Internet: "As computer networks connect the North to the global world, Inuit will continue to adapt media and telecommunications to support Aboriginal culture, language and community; to form Inuit consensus, constituency, institutions and service; to provide education, sustainable resource management and economic development; and to cement their relationships with other circumpolar peoples and other Canadians." Indeed, despite the risks (or perhaps because of them) very few Aboriginal leaders advocate a wholly rejectionist stance toward ICTs, a strategy that would only guarantee that Aboriginal peoples would once again be subjected to a communication system not their own. Instead, most have advocated an approach that stresses the need for Aboriginal peoples to employ ICTs on their own terms. This includes providing a communication infrastructure for indigenous content and cultural practices, for economic and social development in ways that are advantageous and appropriate for Aboriginal communities, for political organization and mobilization, and for a healthy encounter with the non-Aboriginal world. Success stories include Aboriginal use of ICTs to build networks of political empowerment and resistance (Juniper 2002), and to support innovative, decentralized modes of governing in the territorial North (White 2000, 96-7).

Much like the interest groups and social movements on the margins of power discussed in Chapter 4, activists and citizens in Canada's Aboriginal communities have often been more committed to democratic uses of ICTs than have mainstream political institutions and actors. As David Juniper (2002, 149-50) writes in his analysis of Internet activity surrounding the 2000 standoff at the Esgenoopetitj First Nation in New Brunswick, Aboriginal use of the web is "an excellent example of the Internet as a vibrant marketplace of ideas and a dynamic space alive with social, cultural, and political discourse free from state censorship and control." Certainly a significant number of

Aboriginal people and communities have managed to use ICTs in ways that can be described only as empowering. Nevertheless, there is still some distance to travel between this point and the conclusion that these technologies have somehow mediated a more comprehensive redistribution of power between Aboriginal Canadians and the non-Aboriginal state, society, and economy to which they largely remain subject.

The situations of women and Aboriginal people in Canada are symptomatic of the broader politics of the digital divide in this country. Their situations demonstrate that while connectivity is certainly important from a democratic perspective, connectivity alone does not ensure equality relative to ICTs. Numerous other conditions need to be satisfied in order for ICT access and use to contribute to a democratic redistribution of power in Canadian society. As suggested above, these can be grouped loosely into the categories of individual and community capacity to use ICTs effectively, and the ability to determine, or at least influence, the shape of their encounter with these technologies and the ends and limits of their application. Understood this way, the digital divide names not only the gap between those who have access to the Internet and those who do not, but also the difference between those for whom ICTs are an instrument of power and autonomy, and those for whom digital technologies mark a continuing lack of the same.

## The Political Economy of ICTs

A genuinely democratic society will be resolute in separating effective political power from material wealth, social privilege, prestige, and other forms or sources of systemic and prejudicial advantage or disadvantage. The particular genius of a democratic order is that it ought specifically to undermine social and economic inequalities by distributing political power in ignorance of, rather than deference to, them. It is important to note the reference here to the distribution of power, and not just rights. A society that gives every citizen the equal right to vote, express dissent, consult her government, and seek office as a matter of law, but that insulates effective political decision making and action from those processes as a matter of practice (perhaps because decision making responds disproportionately to powerful economic actors) is not a democracy. Democracy denotes a constitutional order, but every constitution has a material foundation, an economy.

An economy that distributes the material resources of effective citizenship relatively equally is foundational to democracy. When material wealth translates into unequal political power, democracy is offended; it is also offended when material disadvantage prevents people from exercising their political capacities effectively as citizens. Two of the most important resources necessary for the practice of citizenship are material security and leisure. In order to engage in public-spirited deliberation over the common good, citizens must be free from the sort of serious material insecurity that quite naturally leads to an overriding concern with one's own self-interest. Although material security provides no antidote to selfishness and does not necessarily translate into public-spiritedness, material insecurity more or less guarantees its impossibility. Indeed, security alone is not really sufficient to provide a material basis for citizenship. For this, leisure is required. Leisure, in this view, denotes time free from the necessities of survival. It should not be confused with recreation, those activities in which we are re-creating ourselves as ready to produce and consume and so also attending to necessity. Citizenship requires time liberated both from the obvious necessity of working to survive and the necessity of recreation to survive work. An economy that fails to distribute the practical resource of leisure relatively equally cannot serve as a material basis for a democracy, because it leaves most people without the time or inclination to engage in citizenship. A crucial mark of a society in which leisure is maldistributed is the professionalization of political life, in which the only people capable of exercising citizenship are those for whom it is also paid work.

An economy that distributes security and leisure relatively equally is a condition of a democratic political order, because these resources are indispensable to widespread participation and inclusion in political life. Digital technology is intimately involved in the distribution of these resources. As much as they are implicated in the elaboration of contemporary practices of politics and governance, digital technologies have also been central to the reconfiguration of the Canadian and global economies. These economies have also had a determining influence on the manner in which these technologies have developed. Indeed, notwithstanding the emerging promise of nonproprietary software coding, or the threat to the music industry posed by peer-to-peer file swapping, the development of these technologies cannot be effectively separated from the realties of contemporary capitalism. Capitalism, it should be remembered, is an economic system *designed* to enable an unequal distribution of material resources and private wealth – it is all about allowing some people to have more security and leisure than others (or, perhaps more accurately, at the expense of others). If the distribution of material security and leisure are foundational to the distribution of power, and if ICTs are implicated in the former and affected by the latter, then the political economy of these technologies is precisely a democratic question (Winseck 2002a).

In Chapters 2 and 3, I presented evidence suggesting that the development and regulation of ICTs has been driven by the priorities of the domestic and global capitalist economy and, more specifically, by powerful corporate actors within those economies. This marks a democratic deficit for two reasons. First, everyday citizens and public interest groups are formally excluded from participation in policy development and decision making. Second, economic power itself is not distributed equally in capitalist societies, and while powerful economic actors are sometimes responsive to consumer and market demands, they are rarely constrained by the deliberation of citizens. From this perspective, the development of ICTs has been substantially determined by a fundamentally undemocratic distribution of power, and has also been instrumental to its entrenchment on a global scale.

If the question is whether ICTs have somehow mediated a radical redistribution of economic power in capitalist societies such as Canada, the answer is surely no. As numerous commentators have observed, digital technology in all its manifestations has been instrumental in the consolidation and concentration of economic power, both domestically and globally (Murdock and Golding 2001; Barney 2000; Dyer-Witheford 1999; Schiller 1999). As discussed in Chapters 2 and 3, this dynamic has been particularly pronounced in the telecommunication and mass media sectors where, despite the reported deceleration of various convergence strategies, levels of global and domestic ownership consolidation have been nothing short of astonishing (Winseck 2002a; Hannigan 2001). And while some might argue that media ownership is irrelevant to content, our estimation of the democratic potential of ICTs must be affected by such an intimate relationship between economic power and media power. As Dwayne Winseck (2002b, 810) puts it: "The bottom line is that communication technologies are being designed more to defend the investment of multinational conglomerates than to further the goal of creating open and transparent mediascapes."

The political economy of ICTs has also affected the distribution of security and leisure. For most Canadians, material security and leisure are a function of their employment situation and work arrangements. Digital technologies have been central to a fundamental restructuring of work and employment in the neoliberal, post-Fordist economies of the so-called network or knowledge societies of the advanced capitalist West (Barney 2004, 91-103; Barney 2000, 132-63; Menzies 1996). Briefly, this restructuring involves (among other things) the rise to prominence of a technologically mediated service economy, the growth of network forms of organization within and between firms, mass customization and just-in-time production and distribution systems, and an overall emphasis on the priority of "flexibility," defined as the ability to adapt quickly to shorter demand cycles, technological innovation, evolving risks, and changing market conditions.

In relation to employment and work, this flexibility has taken the form of chronic destabilization of employment at the individual level, and the growth of nonstandard work arrangements. While there is considerable debate over whether the shift to a knowledge-based, high-tech economy has resulted in net gains or net losses in overall levels of employment in Canada, at the individual level employment is far less stable and predictable than it once was (Aronowitz and Cutler 1998). Due to constant organizational restructuring and technological change, jobs now continually disappear, either permanently or temporarily, before reappearing in a reconstituted form elsewhere. Recurring employment changes have become a routine feature of working life for an increasing number of Canadians. Even if overall levels of employment have improved, the fact remains that many Canadians experience employment as unstable, as a result of the structural characteristics of the labour market in a technological economy.

The character of work itself has also changed (Crow and Longford 2000, 211-16). Standard work once meant a full-time, permanent job completed at regular hours on a relatively fixed schedule, at a place of employment maintained by the employer. Today, nonstandard work – part-time, contract, freelance, temporary, self-employed work, performed at irregular hours on a flexible schedule, often mediated by digital technology off-site from the firm – has almost become a new standard. In Canada, nonstandard forms of employment accounted for 50 percent of all new jobs created in the 1980s and by the 1990s represented over one-third of all employment (Vosko, Zukewich, and Cranford 2003; Crow and Longford 2000, 214). For some, nonstandard employment constitutes empowerment: those for whom part-time and irregular work is a choice and source of independence, rather than their only option, and those for whom self-employment means that they are their own boss rather than a worker without a regular employer.

For an increasing number of people who fit into this category, however, nonstandard work means contingent, precarious, or insecure employment, and it is a condition they have not chosen. Predictably, those who suffer the most under nonstandard working arrangements are those who already occupy positions of significant disadvantage within Canadian society: women, recent immigrants, the under-educated, and the lower working class (Vosko, Zukewich, and Cranford 2003; Vosko 2000). For these people, nonstandard work provides no real relief from insecurity, not only because it cannot be presumed to last, but also because it is inadequately protected by law, and dispensed by employers for whom flexibility means only minimal obligation and contribution to the welfare of their part-time and temporary employees (Fudge, Tucker, and Vosko 2002).

New ICTs have been instrumental to both the destabilization of employment and the rise of nonstandard work arrangements in Canada. Innovation, typically driven or mediated by ICTs, is related structurally to chronic instability in the labour market and to the uneven distribution of employment. And a great deal of work done under nonstandard arrangements is mediated, enabled, coordinated, managed or supervised by ICTs in one form or another (Johnson 2002; Gurstein 2001; Menzies 1997). To the extent they have provided the infrastructure for these conditions, we can reasonably say that ICTs have participated in undermining the material foundations of democratic citizenship. There is a digital divide of sorts at work here, too. For those who suffer at the anxious end of incessant employment instability, involuntary nonstandard work arrangements provide neither the material security nor the leisure to participate in political life as public-spirited citizens. Such people are too busy worrying about their next contract, working extended and unpredictable hours, continually upgrading their skills to keep pace with technical innovation, and basically attending to the needs of their survival (including recreating to endure this cycle). Their material situation condemns them to the privacy of self-interest and alienation (Sennett 1998). This, too, is a consequence of digital technology for democratic citizenship.

Long before the Internet, in response to Plato's utopian dream of communal economic arrangements among the ruling class that would liberate women for participation in public life, Aristotle asked famously, "Who will see to the house?" (Aristotle 1995, 1264b). Here, he expressed not just an indefensible sexism but also an appreciation of the material fact that citizenship becomes possible only when people are released from attention to necessity into leisure, from domestic anxiety into public liberty. In order to herald the possibility of engaged citizenship in Canada, ICTs would have to make a demonstrable contribution to a more equal distribution of material security and leisure in this country. There is considerable reason to believe they have instead reinforced the polarization and inequality between those who enjoy these goods and those who do not. Coupled with the manner in which ICTs have been involved in the consolidation of economic power in Canada more generally, it is hard to imagine how the political economy of these technologies will provide for radical democratic outcomes any time soon.

## The Public Sphere and a Culture of Citizenship

A society that is formally democratic and distributes the material requirements of citizenship equally is still only *really* a democracy if a significant number of its members actually engage in the practices of democratic citizenship on something approaching a regular basis. It makes no sense to say that democracy demands that people participate meaningfully in decision making and action surrounding the common good, and then to affix that label to a society in which most people refrain from doing so most of the time. A democracy is a society in which citizenship is not only possible, but also practised habitually. That is to say, one of the requirements of democracy is a culture of democratic citizenship.

In liberal societies like Canada, we are born with individual rights against our political community but without obligations to participate in the decisions that direct its authority. We are born into *liberal* citizenship (understood as the bearing of rights) but not necessarily as *democratic* citizens (understood as participants in political life). Liberal citizens are born, but democratic citizenship must be cultivated. Citizens are the bacteria of politics: they grow in cultures that nurture them. For a democracy to merit its name it must at least attempt to support a culture that nurtures democratic citizenship and habituates people to its practice. A society whose culture habituates its members to self-interested privatism, individuated pleasure-seeking, consumerism, or cynicism (to name but a few possibilities) in place of democratic citizenship has only the most tenuous claim to being a democracy.

Numerous institutions and factors contribute to generating and sustaining (or undermining) a democratic culture, and not all of these are under scrutiny here. Mass technologies of information and communication, however, are typically singled out as playing a crucial role in this process. Indeed, it is a staple of critical social discourse that mass commercial media tend to enervate the habits of citizenship rather than tending to elevate them. The question in the present context is whether new mass technologies of information and communication contribute to generating and maintaining a culture of citizenship in Canada, or undermine its prospects.

Furthermore, democracy requires not just a culture of citizenship, but also an arena in which it can be exercised. This arena is the public sphere. Since the time of the democratic *polis* in ancient Athens, the public sphere – the sphere beyond the private household – has been understood as a site defined in its *publicness* by democratic citizenship. In his influential account of the rise and fall of the modern public sphere, German political philosopher Jürgen Habermas (1991) described it as a communicative space in which private citizens (i.e., not elected or other officials) together, as equals, engage in debate about the exercise of political authority. To qualify as "public," this sphere must be equally accessible by all citizens, it must be free of arbitrary domination and manipulation, it must feature rational, critical dialogue about the common interest conducted according to shared rules, and it must be not be sacrificed to the economic activities of commerce and consumption. Under these conditions, the public sphere constitutes an arena in which democratic citizenship can be exercised and is a crucial site for demanding that sovereign political power be exercised in a manner responsive to those who authorize it.

Information and communication technologies play a crucial role in mediating the public sphere, such as it exists, in large-scale modern democracies. As Habermas (2001, 102) has pointed out, "When the public is large, this kind of communication requires certain means of dissemination and influence; today, newspapers and periodicals, radio and television are the media of the public sphere." The public sphere of political communication in large-scale democracies is necessarily a mediated sphere, but contemporary mass media do not necessarily fulfill this role adequately. Because of their commercial orientation, and the role they play in manufacturing and manipulating consumption and consent, contemporary mass media tend to act as a "gate through which privileged private interests invade the public sphere" rather than as instruments of rational, critical, political debate among citizens (Habermas 1991, 185). Nevertheless, mass media remain indispensable to the possibility of democratic public spheres organized on the scale of modern countries such as Canada.

A democracy cannot exist unless it maintains a public sphere given over to rational deliberation upon political matters by citizens engaged as equals. That is to say, democracy requires for its functioning a politicized public sphere of freely exercised citizenship. A society in which political deliberation is conspicuous by its relative absence from public life lacks a crucial requirement of a healthy democracy. If the public sphere is exhausted by activities – such as, for example, employment, consumption, and recreation – that leave little or no room for citizenship, then it is difficult to describe that sphere as substantially democratic. The question, in the context of this audit, is whether the proliferation and deployment of ICTs contribute to making the public sphere in Canada one that is characterized by, and hospitable to, the practice of democratic politics.

In terms of the promotion or mediation of a culture of citizenship, it is difficult to find evidence that ICTs are having this effect to a significant degree. Like other Western liberal democracies in the late twentieth and early twenty-first centuries, Canada suffers from a political culture marked by high degrees of alienation, disengagement, and depoliticization. For too many Canadians, citizenship amounts to a claim of rights and entitlements in relation to the state, and to being a passive audience before the spectacle of politics practised in distant places and represented on the various screens that deliver our information and entertainment. It does not include routine participation in the practices of public deliberation, judgment, and action that together define a more substantially political conception of democratic citizenship. As the Canadian Democratic Audit volume on citizens demonstrates, this dynamic of depoliticization is particularly pronounced among young Canadians (presumably the most wired generation). Other, less traditional forms of political action do not in fact seem to compensate for youth disaffection with formal party and electoral politics (Gidengil et al. 2004, 18-40). Politics, it would seem, is just not very high on the list of a Canadian citizen's typical activities.

For the most part, then, a culture of citizenship does not exist in Canada, except in small pockets of politicization: the partisan and policy communities, advocacy groups and social movements, and independent political junkies. The reasons for this are complex and beyond the scope of this investigation. The point is that ICTs exist within the context of a well-established culture of depoliticization in Canada, and it would be the height of technological determinism to assume that these devices might somehow independently reverse this condition. Similarly, we cannot blame ICTs for this depoliticization. Still, it is not unreasonable to ask what sort of contribution ICTs, and the conditions of their development and use, have made to the cultivation of citizenship in Canada. As discussed in Chapter 4, ICTs are an important instrument for those inclined toward the routine practice of democratic citizenship, insofar as they mediate significantly improved access to political information and communication, and provide an infrastructure for the mobilization and coordination of collective political action. Nevertheless, as Chapter 4 also showed, these activities are not characteristic of how the Internet is used by most people most of the time. Instead, contemporary use patterns exhibit a relative deficit of distinctly political uses of the medium by individuals, especially on a widespread or routine basis, and so seem to confirm rather than contest the lack of a well-established culture of citizenship surrounding these technologies.

We also know, from Chapters 2 and 3, that the development of these technologies itself has not consistently been presented as an issue for public, political consideration, and that democratic citizen engagement with ICT design, policy, and regulation has been actively discouraged. And there are no signs that the depoliticization that has been enforced upon technological development in Canada will be reversed any time soon. In technological societies such as Canada, we are culturally predisposed toward regarding the politicization of technological development as something to be avoided, lest it hamper progress or competitiveness in global markets. As Canada's recently articulated "innovation strategy" makes clear, ensuring the country's competitive economic advantage demands a culture of innovation that is supportive of technological invention and risk taking (Industry Canada 2001). Indeed, technologists and politicians alike have raised the issue of cultivating a friendly disposition toward innovation to the level of a national imperative. As a recent report by the association representing Canada's ICT industry puts it: "To build a knowledge-based society we must build a culture of innovation: that means innovation in every aspect of the way we live, work and play" (ITAC 2001, 2). Needless to say, a "culture of innovation" does not have much tolerance for the sort of obstacles that might arise from politicizing technological development, or for the unpredictability of serious democratic deliberation and judgment exercised upon technological change.

Even absent a culture of citizenship, we might ask whether ICTs have contributed to the establishment and maintenance of a democratic public sphere in Canada. The notion of a public sphere in cyberspace is a hotly contested issue in the literature surrounding the politics of digital technology. The idea of the Internet as the site or instrument of a democratic revolution has long been part of the mythology of the medium (Mosco 2004; Barney 2000). Even many critical, progressive intellectuals, however, find good reason to believe that ICTs might produce "new public spheres and spaces for information, debate and participation that contain both the potential to invigorate democracy and to increase the dissemination of critical, progressive ideas" (Kellner 1997, 174), and that the Internet might provide a "supporting foundation upon which public spheres can be built" (Salter 2003, 136).

Much of this hope is based on recognition of the manner in which the technicalities of digital networks – their decentralized architecture and apparent resistance to centralized control, their capacity to mediate both publication and dialogue, their interactive character – seem to overcome many of the liabilities that compromised the ability of previous mass communication technologies to mediate a genuinely democratic public sphere. Some observers have even gone so far as to assert that the Internet constitutes a *radically* democratic, postmodern public sphere, in which individuals are liberated from the arbitrary identity- and geography-based constraints of the off-line world, and in which they can engage in forms of action (hacking, cyber-squatting, identity play, etc.) that, while perhaps not conforming to standard notions of rational dialogue, are nevertheless distinctly political (Poster 2001, 171-88).

Conversely, several critics are skeptical about the potential of ICTs as media of a democratic public sphere. First of all, as discussed earlier in this chapter, access to the Internet in Canada and elsewhere is far from universal, and even those who are connected are not equal users of the medium. Far from being set aside in the on-line world, status differentials continue to pertain among Internet users. Those individuals, groups, and institutions that typically enjoy positions of relative power in Canadian society are also able to use this medium to greater political effect and influence than those classes of users who are also otherwise disadvantaged. Internet access does not make my family the equal of the Asper family, or my friend's anarchist action group the on-line equal of the Liberal Party of Canada.

Some have suggested that the commercial and consumerist priorities driving the design, implementation, and use of the Internet in the period of its maturation and mainstreaming as a communication medium prevent it from fulfilling the conditions of a democratic, political public sphere (Barney 2003; Dahlberg 2003). A public sphere largely given over to the consumption of goods and services (even in the form of information), recreation, idle chatter, hobbyism, and various private entertainments, and in which political discourse is a marginal activity at best, is more like a market than a democratic political space. But the criticism refers not just to the use of the Internet as an infrastructure for commerce: the norms of discourse on the Internet do not characteristically meet the requirements of rational critical debate aimed at reaching reasonable consensus on complex, contestable issues where contending interests are in play (Dahlberg 2001). In a similar vein, Hubert Dreyfus (2001) maintains that the anonymity enabled by network-mediated communication, apparently so conducive to the proliferation of free expression of opinion, actually undermines the experience of commitment, risk, and responsibility that makes political action in the public sphere meaningful.

Additionally, several critics have argued that, even if the Internet mediates something like a public sphere in which political dialogue occurs among citizens, the customization enabled by digital media means that the political space will be fragmented into self-selecting and mutually exclusive enclaves. In this view, the Internet comprises not one public sphere but rather many public *sphericules*, in which like-minded and similarly interested people meet to reinforce, rather than negotiate, their private preferences and to avoid, rather than encounter, the differences that characterize a pluralistic society such as Canada's (Sunstein 2001; Gitlin 1998).

Yet another reason to be critical of the relationship between ICTs and the democratic public sphere arises from the fact that digital technologies have been instrumental in an escalation of contemporary practices of surveillance. Powerful actors and institutions have used digital networks to engage in intensive and extensive monitoring of activities on-line and everyday behaviour off-line (Lyon 2003; Whitaker 1999). These practices range from systematic collection, storage, integration, and processing of digitized information identified with individuals, to the incorporation of physical surveillance systems (e.g., closed circuit cameras, digitized building security systems, and biometric devices) into more comprehensive digital information systems. For the most part, concern about the escalation of electronic surveillance has been expressed in terms of the heightened risks these practices and technologies pose for individual privacy and the security of personal information, a concern that, for example, inspired Canada's Personal Information Protection and Electronic Documents Act, first passed in 2001 and fully operational in 2004.

Recently, however, concern has been growing over the negative impact of increased electronic surveillance not only on individual privacy, but also on the democratic public sphere. It is important to recall, in this context, that ideally the public sphere is a space that has not been colonized by the forces of the state or the market, a sphere in which individual citizens can interact freely as equals without fear of domination and manipulation by powerful economic and state actors. Of course, the recent escalation of digitally mediated surveillance has been driven by both the state and private corporations. Both present increased surveillance as necessary to calculate and manage the myriad risks that supposedly confront states and markets in the contemporary period, risks that might otherwise undermine values ranging from state and personal security, to corporate profitability and competitiveness.

Surveillance is also often presented as a necessary adjunct to consumer convenience, a price we pay to be able to access and use digital technologies at all. As David Lyon (2003, 10) has put it, contemporary surveillance is about creating "categories of suspicion" as well as "categories of seduction." Of course, the risks that increased commercial and state surveillance pose for the exercise of democratic citizenship rarely enter into this calculus. Nevertheless, digitally enhanced surveillance has become perhaps the chief means by which powerful state and economic actors have occupied the public sphere, converting it from a space of unencumbered discussion among equal citizens into a space characterized by incessant supervision, generalized suspicion, and disciplinary social categorization. Somehow, the same technology that was supposed to produce a virtual democratic agora has helped to give the entire experience of being in the public sphere the feeling of travelling through a major international airport.

This is not to suggest that because of digital technology, liberal democracies such as Canada have suddenly become police states. Rather, when state and commercial actors use digital technologies to construct the public sphere as a space in which citizens going about their daily business are subjected to extensive and intensive surveillance, the consequences for democratic politics are quite serious. The problem is not so much that the Canadian state has or will become brutally oppressive of civil liberties, but that the subjection of citizens to routine surveillance in a climate of suspicion has a chilling effect on otherwise perfectly legitimate civic activity, and so fundamentally depoliticizes the public sphere. As Lyon (2003, 44-5) analyzed the escalation of surveillance since the 2001 attacks on the World Trade Center in New York, "Something deeply anti-political imbues the new security measures ... political participation [is] being eclipsed by risk management." This is not only because widespread surveillance might dissuade people from political action they would otherwise have freely undertaken. It is also because when automated, technologically mediated surveillance and risk management based on computer-generated profiles become a de facto mode of governance – a means by which it is decided who receives benefits, privileges, protection, or punishment, and who does not – the role of democratically exercised political judgment is systematically diminished. In this respect, the role ICTs play in contemporary surveillance regimes constitutes another digital divide between those for whom these technologies are a source of power and autonomy and those for whom they are not.

## ICTs and the Distribution of Power

This chapter began by pointing out that democracy is really about the distribution of power, that its defining principle is equality, and that the realization of this principle in practice relies as much upon material conditions as upon formal constitutional arrangements. In other words, the question of whether the Canadian political system is adequately inclusive, participatory, and responsive is ultimately a question about whether material and political power in Canadian society is distributed more or less equally. It was in this light that we considered the relationship between ICTs and the distribution of power in Canadian society, both in terms of how these technologies have affected this distribution and how existing power relations have influenced the development and democratic potential of these technologies.

Little evidence was found to support the claim that ICTs have been involved in a fundamental redistribution of power in Canada; considerably more was found to suggest that ICTs both reflect and reinforce the existing inequalities. The digital divide – whether conceived narrowly in terms of connectivity, or more broadly in terms of the ability and capacity to use ICTs in ways that enhance the autonomy of disadvantaged citizens – is an enduring reality in Canada's corner of cyberspace. Similarly, I argued that ICTs have been a central instrument in the consolidation of economic and media power in Canada, and have done little to redress the unequal distribution of material security and leisure, resources that make active democratic citizenship possible. Finally, I found little reason to believe that ICTs are involved in a rebirth of the culture of democratic citizenship in Canada, or that the relationship between ICTs and the prospect of genuinely democratic public sphere in Canada is anything more than ambiguous, at best.

None of this appears to bode well for the possibility of new information technologies mediating a more responsive, inclusive, and participatory political life in Canada; it seems to suggest that ICTs have been part of tendency away from, rather than toward, enhanced democracy. Be that as it may, it bears repeating that none of what is described above is a necessary outcome of the technical character of ICTs themselves. The development and deployment of these technologies could be limited by the principle of equality, and they can be enlisted in its service. The various digital divides discussed above are less a product of digital technology than they are a creature of the political economy of Canada. Any hope for democratic renewal in Canada should therefore rely more heavily on the difficult struggle for socioeconomic justice than on easy technological fixes.

## Chapter 5

### Strengths

- Overcoming access divides is consistently expressed as a goal of ICT policy in Canada.
- Some communities, including Aboriginal communities, have used ICTs in innovative ways despite the persistence of access barriers.
- Many Canadians use ICTs in ways that enhance their autonomy and flexibility, and increase their citizenship resources.

### Weaknesses

- Power in relation to ICTs is unevenly distributed in Canadian society, typically in ways that mirror existing axes of systemic inequality and disadvantage.
- Many Canadians experience ICT-mediated restructuring as a source of insecurity that undermines their capacities as citizens.
- New ICTs have been instrumental to privatization and increased surveillance in the public sphere in Canada.

# 6

# THE QUESTION

Whenever I tell people that I study the politics of digital technology, the first question they typically ask is some variation of "So, is the Internet good for democracy or isn't it?" Academics are not as agile in evading questions as politicians are, but we're not bad at it either. And so I am usually able to come up with some sort of non-answer that gets me out of the conversation: "It's too soon to tell"; "That's a big question!"; "It's kinda complicated"; "Well, now, that depends on what you mean by 'democracy.'" I am never sure whether people are disappointed by these responses or comforted by them.

The Canadian Democratic Audit has asked us to consider whether the growing presence of new information and communication technologies in our midst has helped to make politics in Canada more democratic – specifically, more inclusive, participatory, and responsive – than it was prior to their arrival. In the preceding pages, I have not said much to indicate an affirmative answer to this question. First, we looked at the development of these technologies as an object of political judgment and a matter of public policy, and found that, in a marked departure from the history of Canadian communication and cultural policy, the development of ICTs in Canada has largely been exempted from democratic consideration and priorities. Next, we considered the relationship between these technologies and the dynamics of capitalist globalization, which make the democratic exercise of sovereignty at the national or local level so very difficult to sustain. I also questioned the efficacy of technological-nationalist responses to this quandary – responses that risk depoliticizing technological development even further, and that are enlisted more easily to bolster the competitive advantage of Canadian capitalists than to enhance the democratic influence of Canadian citizens.

Following this, we examined the particularities of ICT use by the major institutions and actors that make up the political scene in Canada: government, political parties, advocacy groups and social movements, and citizens. ICTs have certainly had an impact upon the informational and communicative aspects of political life in Canada. But examples of the considerable potential of these instruments being fulfilled to mediate more inclusive, participatory, and responsive civic practices were relatively few, and quite marginal both to the mainstream, institutional exercise of political authority in this country, and to the way most people use these technologies most of the time. Finally, we investigated the relationship between these technologies and the distribution of political and economic power in Canada, and the manner in which they have affected the material conditions, culture, and public sphere in which democratic citizenship is exercised. Little evidence was found to suggest these technologies have been involved in an equalization of power in Canada, a redistribution of the civic resources of material security and leisure, support for a culture of citizenship, or a significant democratization and politicization of the public sphere.

And so I return to that nagging question to which, I suppose, this book is a rather long answer. Admittedly, the words participation, responsiveness, and inclusiveness are not the first that come to mind when I think about the politics of ICTs. That being said, we must also recognize the democratic opportunities before us and encourage the possibility of seizing them.

With regard to ICT policy, Canada has a significant history of widespread public consultation in communication and cultural policy, especially whenever technological change recommends reconsideration of established assumptions, priorities, and instruments. It is precisely this history of public engagement – and the actors who carried it out – that we have to thank for many of the most characteristically democratic elements of the Canadian media and communication system. As documented in Chapter 2, the ICT policy cycle has departed from this tradition, largely because capitalist elites in Canada have managed to position their private interests in a manner that leaves little room for democratic involvement in communication and technology policy. But it does not have to be this way. The other thing that Canada's particular history with information and communication technology has produced is a highly mobilized, literate, and passionate community of advocates around these issues, a community that has led the struggle for the public interest in communication and cultural policy. This community has been demoralized by the turn away from democracy in the ICT policy cycle, but it could easily be rejuvenated by a genuine turn back toward democracy. The venues that still exist for public participation – such as the CRTC, which is set to review its approach to new media in the near future – must be committed once again to processes that are public and transparent, participatory and inclusive, and perhaps most importantly, genuinely responsive to the range of interests expressed surrounding these issues.

Beyond this, the democratic mistakes that characterized priority setting and policy making before and during the building of Canada's ICT infrastructure must not be allowed to persist through subsequent stages of its development. The closed-door, industry-dominated approach that has emerged under Industry Canada's leadership must give way to institutional opportunities for participatory, inclusive, and responsive engagement of the public interest, on a scale that lives up to, rather than runs away from, Canadian tradition in this crucial area. At a national level, this might include the establishment of a *Public* Information Highway Advisory Council, an institutionalized version of the ad hoc network of advocacy groups that gathered under this name upon being excluded from the original IHAC process in the

1990s. Such a group might do for the distinctly public interest in ICTs what IHAC did for private interests, namely, provide a forum for identifying challenges and opportunities, establishing priorities, and devising action strategies and accountability mechanisms in specific relation to ICT-related public goods. Such goods might include, for example, the following:

- the right to communicate unmolested by corporate control
- the integrity of noncommercialized public spheres
- deeply egalitarian access to the infrastructures and capacities of public information and communication
- limits on the concentration of mass media ownership
- adequately resourced community networks
- promotion of culturally diverse content
- privacy protection beyond that which is required to make the world safe for electronic commerce.

They might also include the possibility of not developing certain technologies at all, or not allowing current technology to be developed in particular directions, should such restraint be warranted by the public interest. Whatever it might come up with, a Public Information Highway Advisory Council would also have to be more participatory and inclusive than the exclusive IHAC, which offered almost no opportunity for public participation, and whose imagination was consequently consumed by the technological imperatives and neoliberal rhetoric advanced by the industry representatives who dominated the council.

Similar things might be said of the relationship among democracy, cultural and political sovereignty, and global markets and media technologies. Chapter 3 detailed the very real challenges that globalization poses for Canadian democracy, and also the intimate relationship between these challenges and ICTs. But globalization is not a disembodied historical force immune to intervention, and its antidemocratic version is not our destiny. Much can be done to recover democratic agency in the context of globalization.

With its state partners in the International Network on Cultural Policy, and in multilateral agencies such as UNESCO and the International Telecommunication Union (ITU), Canada should play a leading role in continuing to press for new international instruments and agreements that protect culture and communication from the ravages of borderless markets, in which a handful of transnational media conglomerates dominate communication and obscure their exemption from social obligation with the rhetoric of innovation, convergence, and competitiveness. Of course, in order to do so, the Canadian state must make a material commitment to a communication and cultural policy driven by public interest priorities, even if this means becoming less competitive in the global race for international investment and transnational commerce. Should it make such a commitment, Canada could exert leadership by advocating radical democratization of the venues and institutions in which global media, cultural, and communication policy is made, such as the WTO, the OECD, the G8, and their related agencies.

This would include pushing for greater transparency, effective civil society representation, and enforceable accountability to the various publics who live in the world made by global capital, its technologies, and its governing institutions. The prospect of an even somewhat democratized global order cannot withstand *in camera* negotiations, in which only the representatives of powerful states and global business conglomerates are invited to participate. Accordingly, Canada should press the ITU to accelerate its recognition of transnational civil society representatives as equal partners alongside states, UN agencies, and the private sector in forums such as the World Summit on the Information Society (WSIS) (Moll and Shade 2004a). Well-established nongovernmental organizations such as, for example, the Association for Progressive Communications (and its affiliated Women's Networking Support Program), Communication Rights in the Information Society, and the Platform for Community Networks must be allowed a seat at the table, from which they can voice and defend the globally shared but locally defined public interest in democratic communication.

In this respect, the Canadian state should strenuously support the work of Canadian activists, researchers, and organizations that are active domestically and that participate in transnational civil society networks surrounding ICT-related issues and globalization more generally. These activists struggle against long odds to establish things like an international right to communicate, and the right for local communities to set the terms of their own encounters with ICTs, as minimum conditions of democratic legitimacy in the global technological order. Serious state support could shorten these odds considerably. Such support could range from providing public funding and infrastructural support to these actors, to actively consulting them prior to engagement in international venues where decisions about global communication are made.

To date, the Canadian state has failed on both accounts. Civil society groups eager to participate in the first phase of the WSIS process received no financial support from the Canadian state, leading one long-time activist and scholar in this area to observe that "the Canadian presence at the WSIS has for the most part consisted of presentation of (self-serving) governmental 'success stories' rather than the rather more useful and authentic experiences of those who have been working for years in the field in these areas" (Michael Gurstein, quoted in Moll and Shade 2004a, 62). In the lead-up to its own participation in the first phase of WSIS, Industry Canada did solicit input through a selectively mailed brochure and a website questionnaire, both requiring response on a short deadline. Questions were either limited in scope to Industry Canada's pet concerns and vocabulary (e.g., partnerships, best practices, etc.) or leading (e.g., "How have ICTs contributed to community and quality of life?"). A total of sixteen people responded to the website. And, as Moll and Shade observe, "There is no attempt made to consolidate or analyze those views. There is no indication on the site about how those views were incorporated into the government's position" (64). The community of ICT activists in Canada is extensive, motivated, and vociferous. That Industry Canada's call for input drew so few responses is symptomatic of the deep malaise that has afflicted its leadership of Canadian domestic and international

ICT policy: a growing, almost wilful, divide between its core concerns and those of civil society actors animated by the public interest in ICTs; and Industry Canada's consistent failure to respond to the needs, priorities, and ideas of this constituency. In fact, these public interest and community advocates seem to have realized that they are not Industry Canada's constituency at all. This will have to change if Canada wishes to show genuine leadership in the democratization of global ICT policy.

As to the prospects of ICT use by democratic institutions and actors, as suggested in Chapter 4, much depends on the motivations driving adoption and deployment of these technologies. Many advocacy groups and new social movements have been successful in using ICTs in ways that are responsive, inclusive, and participatory, because genuine democratic engagement is an important part of their identity, and often inseparable from the other goals animating their activity. Political uses of ICTs driven by the pursuit of operational efficiency, public relations management, and legitimation strategies are unlikely ever to attain the standards of participation, inclusiveness, and responsiveness set out in this audit, because these are simply not the goals of such efforts. In order to exploit fully the democratic potential of ICTs, governments and political parties will have to mimic advocacy groups and social movements by committing themselves to genuine democratic engagement as a *primary* goal, come what may. Given the resources at their disposal, these actors and institutions could be as successful in this endeavour as they have been in using ICTs for purposes of service delivery, public relations, and strategic information gathering and processing. Nor should this be too great a stretch for them, as their own rhetoric already points them in this direction.

Finally, there remains the difficult matter of the relationship between ICTs and the distribution of socioeconomic power in Canada, a relationship I characterized in Chapter 5 as undermining the material foundation of participatory, responsive, and inclusive democratic practice. It is tempting to say that nothing short of a radical restructuring of capitalism in Canada, aimed at reducing systematic material inequalities and eliminating their correspondence to the unequal distribution of political power in this country, will produce an environment in which ICTs can be salvaged for substantially democratic application. But that would be too easy. More difficult is a sober assessment of what can be done within the horizon of immediate possibility, which might include the following measures:

- the adoption of an expansive, rather than restricted, definition of the digital divide by the Canadian state in its efforts to reach the goal of universal access to ICTs
- an effort to provide increased public financial and regulatory support for the creation and maintenance of noncommercial spaces, such as community networks, on the Internet
- reform of labour laws to protect workers rendered insecure and vulnerable by technological change, including structural changes mediated by digital technology (for example, the proliferation of precarious, home-based employment)
- serious limits on state and commercial surveillance of everyday activity in the digital public sphere.

Such measures will not eliminate the inequality engendered by the capitalist economy, nor will they free ICTs from their relationship to this inequality. They may, however, at least provide some material support for citizenship, and contribute to a social distribution of resources and power that encourages, rather than discourages, its practice.

The answer to the question of whether ICTs are good for democracy is suggested by the priorities and habits of those who control and use these technologies in political life. As I write this, the 2004 general election campaign has come and gone. Sadly, early analysis of ICT use in this election cycle reveals a continued poverty of imagination on the part of Canada's political parties when it comes to these technologies. In a study of candidates' websites in the Conservative Party leadership contest that preceded the election, Jonathan Rose and Tamara Small (2004) concluded that the primary audience for these sites were the mainstream mass media, and that citizen engagement was a secondary concern. This pattern appears to have extended into the election campaign itself. Initial research on the campaign has found that "Canadian sites reflect a top-down command-and-control campaign model. The information flow is largely unidirectional – from the party to the public/supporters ... They offer no substantive means through which party grassroots can organize, mobilize, share practices, download key campaign tools, and coordinate outreach. Canadian sites resemble electronic lawn signs – they inform but don't engage" (Hillwatch 2004, 1-2).

Digital technology certainly played an important role in the campaign. Party strategists wielding hand-held computers connected to wireless networks communicated with a similarly equipped press corps, in rapid-fire efforts to seize the campaign agenda, undermine their adversaries, drive media coverage, and respond instantly to similar tactics employed by opposing campaigns. Digital messages containing accusations, revelations, rebukes, and compromising quotations, images, and video clips circulated from party researchers to hundreds of reporters across the country in an instant. Quickly, the contest assumed the rhythm of firing a volley, ducking to avoid one, and then firing again, with partisans wielding ICTs as they would weapons on a battlefield. Of course, for a couple of decades now, parties have been calling their strategic headquarters "war rooms." But this was a political campaign, not a military one, an election, not a war. War, not to put too fine a point on it, is a symptom of democracy's failure, not its defining event, as we generally assume elections to be.

ICTs might be deployed to elevate the democratic character of elections, rather than debasing them into some sort of militarist spectacle. Sadly, it is difficult to imagine how such deployments might actually take place under contemporary political, economic, cultural, and technological conditions in Canada. Genuine democratic politics takes more imagination, and more courage, than war games, and new weapons are no surrogate for these virtues. Indeed, new weapons can easily seduce us into believing there is no need for imagination and courage, and that better technology will deliver the goods. This, I think, is precisely where we stand with new information and communication technologies: at risk of being persuaded that we have been relieved of the need to be imaginative and courageous in our struggle to realize the democratic principles we hold so dear.

# Discussion Questions

## Chapter 1: Democracy, Technology, and Communication in Canada

1. What does democracy mean to you? Why should democracies be inclusive, participatory, and responsive?
2. How would you characterize the relationship between politics and technology? In what sense(s) is technology political? Is there a special relationship between politics and communication technology in particular?
3. What factors contribute to the political outcomes of technological development? What shape do these factors take in contemporary Canada?

## Chapter 2: The Politics of Communication Technology in Canada

1. Should issues and decisions surrounding technological development in Canada be subjected to democratic deliberation? Why or why not? Does democratic deliberation always lead to good policy outcomes?
2. Using the standards of inclusiveness, participation, and responsiveness, how would you assess the development of ICT policy and regulation in Canada?
3. How could the democratic character of ICT policy and regulation in Canada be improved?

## Chapter 3: Communication Technology, Globalization, and Nationalism in Canada

1. Is a nationalist communication policy good for democracy in Canada? What other priorities should drive communication policy in Canada?
2. What role have ICTs played in the dynamics of economic, political, and cultural globalization? How have these dynamics affected the development of ICTs in Canada?
3. What are the implications of globalization for democracy in Canada? How can inclusiveness, participation, and responsiveness be secured under the pressures of globalization? Can ICTs play a role?

## Chapter 4: Technologies of Political Communication in Canada

1. Are the federal government's efforts in the Government On-Line initiative making Canadian democracy more inclusive, participatory, and responsive? How could these efforts be improved?

2 Has the use of digital technology by Canadian political parties made them more inclusive, participatory, and responsive? What about the use of these technologies by social movements and advocacy groups?
3 What is the potential for ICTs to enhance the everyday democratic experience and practices of citizens? Why haven't citizens made the most of these technologies?

## Chapter 5: Digital Divides

1 What is the relationship between economic equality and democracy? How have ICTs affected (or been affected by) the distribution of power in Canada?
2 What is the digital divide, and why is it a democratic problem?
3 What contribution have ICTs made to the culture of citizenship in Canada? Have these technologies made the public sphere more inclusive, participatory, and politicized?

## Chapter 6: The Question

1 Are ICTs good for democracy in Canada? Give reasons for your answer.

# Additional Reading

### Chapter 1: Democracy, Technology, and Communication in Canada

For theoretical discussion of the general relationship between politics and technology, see Langdon Winner's *The whale and the reactor: A search for limits in an age of high technology* (1986), Ursula Franklin's *The real world of technology* (1999), and Andrew Feenberg's *Questioning technology* (1999). In specific relation to ICTs, see Vincent Mosco's *The digital sublime* (2004). For critical analysis of democracy and ICTs, see Anthony Wilhelm's *Democracy in the digital age* (2000) and my own *Prometheus wired: The hope for democracy in the age of network technology* (2000). In the Canadian context, see the collections edited by Marita Moll and Leslie Regan Shade, *E-commerce vs. e-commons: Communications in the public interest* (2001), and Cynthia Alexander and Leslie Pal, *Digital democracy: Policy and politics in the wired world* (1998).

### Chapter 2: The Politics of Communication Technology in Canada

On the history of communication policy in Canada see the following: Robert Babe's *Telecommunications in Canada: Technology, industry, government* (1990); Marc Raboy's *Missed opportunities: The story of Canada's broadcasting policy* (1990); Michael Dorland's edited collection, *The cultural industries in Canada* (1996); and Dwayne Winseck's *Reconvergence: A political economy of telecommunications in Canada* (1998). On ICT-related policy, see Marita Moll and Leslie Regan Shade's *Seeking convergence in policy and practice: Communications in the public interest,* vol. 2 (2004b), and Donald Gutstein's *E.con: How the Internet undermines democracy* (1999). The Information Policy Research Program in the Faculty of Information Studies at the University of Toronto has archived several useful reports in this area at www.fis.utoronto.ca/research/iprp/publications, such as Clement and Shade 2000 and McDowell and Buchwald 1997.

### Chapter 3: Communication Technology, Globalization, and Nationalism in Canada

The crucial work on technological nationalism in Canada is Maurice Charland's 1986 essay "Technological nationalism." See also Richard Collins's *Culture, communication and national identity: The case of Canadian television* (1990). For analysis of the threat posed by capitalist media to cultural sovereignty in Canada, see Dallas W. Smythe's *Dependency road: Communications, capitalism, consciousness and Canada* (1981). On ICTs and globalization, see Ronald Deibert's *Parchment, printing and hypermedia: Communication in world order transformation* (1997) and the collection *Street protests and fantasy parks: Globalization, culture, and the state,* edited by David R. Cameron and Janice Gross Stein (2002c). On Canadian communication policy in the era of globalization, see the report by the Standing Committee on Canadian Heritage, *Our cultural sovereignty: The second century of Canadian broadcasting* (2003), Bram Abramson and Marc Raboy's article, "Policy globalization and the 'information society': A view from Canada" (1999), and Vincent Mosco and Dan Schiller, eds., *Continental order? Integrating North America for cybercapitalism* (2001).

### Chapter 4: Technologies of Political Communication in Canada

For a comprehensive review of issues surrounding e-government in Canada, see the Centre for Collaborative Government's eleven-volume Crossing Boundaries: Changing Government series, available at www.crossingboundaries.ca. (The series includes Alcock and Lenihan 2001, Crossing Boundaries Political Advisory Committee 2003, Kernaghan, Riehle, and Lo 2003, and Lenihan 2002a, 2002b, and 2002c.) For an appraisal of the federal Government On-Line project, see the 2003 report of the Office of the Auditor General of Canada. For a discussion of the possibilities of e-democracy, see the report *Bowling together: Online public engagement in policy deliberation,* by Stephen Coleman and John Gøtze (2001). On the use of ICTs by governments, parties, advocacy groups, and citizens globally, see Pippa Norris's *Digital divide: Civic engagement, information poverty, and the Internet worldwide* (2001). On interest groups and social movements, see Martha McCaughey and Michael Ayers, eds., *Cyberactivism: Online activism in theory and practice* (2003). In the Canadian context, see Manjunath Pendakur and Roma Harris, eds., *Citizenship and participation in the information age* (2002).

### Chapter 5: Digital Divides

On the digital divide and access strategies in Canada, see Andrew Clement and Leslie Regan Shade's "The access rainbow: Conceptualizing universal access to the information/communication infrastructure" (2000) and Mark Warschauer's *Technology and social inclusion: Rethinking the digital divide* (2002). On gender and ICTs, see Shade's *Gender and community in the social construction of the Internet* (2002) and Melanie Stewart Millar's *Cracking the gender code: Who rules the wired world?* (1998). On Aboriginal people and new media, see Gail Valaskakis's article "Remapping the Canadian north: Nunavut, communications and Inuit participatory development" (2002), David Juniper's "The Moccasin Telegraph goes digital: First Nations and political usage of the Internet" (2002), and Cynthia Alexander's "Wiring the nation! Including First Nations? Aboriginal Canadians and federal e-government initiatives" (2001b). On the political economy of ICTs, see Nick Dyer-Witheford's *Cyber-Marx: Cycles and circuits of struggle in high-technology capitalism* (1999) and Heather Menzies's *Whose brave new world? The information highway and the new economy* (1996). On the impact of ICTs on the distribution of power in the Canadian economy, see Barbara Crow and Graham Longford's essay "Digital restructuring: Gender, class and citizenship in the information society in Canada" (2000). On ICTs and the public sphere, see my "Invasions of publicity: Digital technology and the privatization of the public sphere" (2003).

# Works Cited

## Statutes and Treaties

Broadcasting Act, 1991, c. 11.

Canada Elections Act, 2000, c. 9.

FTA (US-Canada Free Trade Agreement), 1989.

Government Organization Act, 1969.

NAFTA (North American Free Trade Agreement), 1993.

Telecommunications Act, 1993, c. 38.

## Other Sources


Abramson, Bram, and Marc Raboy. 1999. Policy globalization and the "information society": A view from Canada. *Telecommunications Policy* 23: 775-91.

Accenture. 2003. Canada leads 22 countries in developing eGovernment for third consecutive year. Press release. 8 April. <www.newswire.ca/fr/releases/archive/April2003/08/c8818.html> (16 February 2004).

Alcock, Reg, and Donald Lenihan. 2001. *Opening the e-government file: Governing in the 21st century.* Crossing Boundaries Project: Changing Boundaries, vol. 2. Ottawa: Centre for Collaborative Government.

Alexander, Cynthia. 2000. Cents and sensibility: The emergence of e-government in Canada. In *How Ottawa spends,* ed. Leslie A. Pal, 185-209. Don Mills, ON: Oxford University Press Canada.

—. 2001a. Digital leviathan: The emergence of e-politics in Canada. In *Party politics in Canada,* 8th ed., ed. H. Thorburn and A. Whitehorn, 460-76. Toronto: Pearson.

—. 2001b. Wiring the nation! Including First Nations? Aboriginal Canadians and federal e-government initiatives. *Journal of Canadian Studies* 35(4): 277-96.

Alexander, Cynthia, and Leslie Pal, eds. 1998. *Digital democracy: Policy and politics in the wired world.* Don Mills, ON: Oxford University Press Canada.

Anderson, Benedict. 1991. *Imagined communities: Reflections on the origin and spread of nationalism.* London: Verso.

Aristotle. 1995. *Politics,* trans. E. Barker. London: Oxford University Press.

Aronowitz, Stanley, and J. Cutler, eds. 1998. *Post-work: The wages of cybernation.* New York: Routledge.

Attallah, Paul, and Angela Burton. 2001. Television, the Internet, and the Canadian federal election of 2000. In *The Canadian general election of 2000,* ed. Jon Pammett and Christopher Dornan, 215-41. Toronto: Dundurn Press.

Austen, Ian. 1994. Federal advisers on information highway taking a secret route. *Vancouver Sun*, 17 March, D7.

Babe, Robert. 1990. *Telecommunications in Canada: Technology, industry, government.* Toronto: University of Toronto Press.

Balka, Ellen, and Richard Smith, eds. 2000. *Women, work, and computerization: Charting a course to the future.* Norwell, MA: Kluwer.

Barney, Darin. 1996. Pushbutton populism: The Reform Party and the real world of teledemocracy. *Canadian Journal of Communication* 21(3): 381-413.

—. 2000. *Prometheus wired: The hope for democracy in the age of network technology.* Vancouver: UBC Press.

—. 2003. Invasions of publicity: Digital technology and the privatization of the public sphere. In *New perspectives on the public-private divide,* ed. Law Commission of Canada, 94-122. Vancouver: UBC Press.

—. 2004. *The network society.* Cambridge: Polity.

Barney, Darin, and David Laycock. 1999. Right-populists and plebiscitary politics in Canada. *Party Politics* 5(3): 317-40.

Becker, Ted, and Christa Daryl Slaton. 2000. *The future of teledemocracy.* Westport, CT: Praeger.

Bertrand, Françoise. 1999. Notes for an address on the occasion of the release of the *Report on new media.* Canadian Radio-television and Telecommunications Commission. <www.crtc.gc.ca> (August 1999).

Bimber, Bruce. 1998. The Internet and political transformation: Populism, community and accelerated pluralism. *Polity* 31(1): 133-60.

—. 2001. Information and political engagement in America: The search for effects of information technology at the individual level. *Political Research Quarterly* 54(1): 53-67.

Brodie, Janine. 1996. Restructuring and the new citizenship. In *Rethinking restructuring: Gender and change in Canada,* ed. Isabella Bakker, 126-40. Toronto: University of Toronto Press.

Cameron, David R., and Janice Gross Stein. 2002a. The state as a place amidst shifting spaces. In *Street protests and fantasy parks: Globalization, culture and the state,* ed. David R. Cameron and Janice Gross Stein, 141-59. Vancouver: UBC Press.

—. 2002b. Street protests and fantasy parks. In *Street protests and fantasy parks: Globalization, culture and the state,* ed. David R. Cameron and Janice Gross Stein, 1-19. Vancouver: UBC Press.

—, eds. 2002c. *Street protests and fantasy parks: Globalization, culture and the state.* Vancouver: UBC Press.

Canada. 1999. Speech from the throne to open the second session, thirty-sixth parliament of Canada. October 12. Ottawa: Privy Council Office.

—. 2001. Speech from the throne to open the first session, thirty-seventh parliament of Canada. Ottawa: Privy Council Office.

Canadian Election Study. 2000. Post-election questionnaire 2000. <www.fas.umontreal.ca/POL/ces-eec/ces.html> (24 February 2003).

Carty, R. Kenneth. 1988. Three Canadian party systems: An interpretation of the development of national politics. In *Party democracy in Canada: The politics of national party conventions,* ed. George Perlin, 15-31. Scarborough, ON: Prentice-Hall.

Carty, R. Kenneth, William Cross, and Lisa Young. 2000. *Rebuilding Canadian party politics.* Vancouver: UBC Press.

Castells, Manuel. 1997. *The power of identity.* London: Blackwell.

—. 2001. *The Internet galaxy: Reflections on the Internet, business and society.* Oxford: Oxford University Press.

Chadwick, Andrew, and Christopher May. 2003. Interaction between states and citizens in the age of the Internet: E-government in the United States, Britain and the European Union. *Governance* 16(2): 271-300.

Charland, Maurice. 1986. Technological nationalism. *Canadian Journal of Political and Social Theory* 10(1-2): 196-220.

Clarkson, Stephen. 2002. *Uncle Sam and us: Globalization, neoconservatism and the Canadian state.* Toronto: University of Toronto Press.

Clement, Andrew, Marita Moll, and Leslie Regan Shade. 2001. Debating universal access in the Canadian context: The role of public interest organizations. In *E-commerce vs. e-commons: Communications in the public interest,* ed. Marita Moll and Leslie Regan Shade, 23-48. Ottawa: Canadian Centre for Policy Alternatives.

Clement, Andrew, and Leslie Regan Shade. 2000. The access rainbow: conceptualizing universal access to the information/communication infrastructure. Information Policy Research Program working paper no. 10, Faculty of Information Studies, University of Toronto. <www.fis.utoronto.ca/research/iprp/publications> (2 July 2002).

Coleman, Stephen, and John Gøtze. 2001. *Bowling together: Online public engagement in policy deliberation.* London: Hansard Society.

Collins, Richard. 1990. *Culture, communication and national identity: The case of Canadian television.* Toronto: University of Toronto Press.

Commission on the Future of Health Care in Canada. 2002. *Building on values: The future of health care in Canada.* Ottawa: Commission on the Future of Health Care in Canada.

Connecting Aboriginal Canadians Forum. 2003. *Final report.* Ottawa. <www.aboriginalcanada.gc.ca/cac/lookups/cacdocs.nsf/vDownload/april30_w_contact.pdf> (21 March 2004).

Courtney, John C. 2004. *Elections*. Canadian Democratic Audit. Vancouver: UBC Press.

Cross, William. 1998. Teledemocracy: Canadian political parties listening to their constituents. In *Digital democracy: Policy and politics in the wired world,* ed. C. Alexander and L. Pal, 132-48. Don Mills, ON: Oxford University Press Canada.

Crossing Boundaries National Council. 2003. Summaries, Crossing Boundaries National Conference. <www.crossingboundaries.ca> (2 April 2004).

—. 2004. MP website study. Ottawa: Centre for Collaborative Government. <www.crossingboundaries.ca> (2 April 2004).

Crossing Boundaries Political Advisory Committee. 2003. *Finding our digital voice: Governing in the information age.* Crossing Boundaries Project: Changing Government, vol. 11. Ottawa: Centre for Collaborative Government.

Crow, Barbara, and Graham Longford. 2000. Digital restructuring: gender, class and citizenship in the information society in Canada. *Citizenship Studies* 4(2): 207-30.

Crowley, David. 2002. Who are we now? Contours of the Internet in Canada. *Canadian Journal of Communication* 27(4): 469-507.

CRTC (Canadian Radio-television and Telecommunications Commission). 1995. *Competition and culture on Canada's information highway: Managing the realities of transition.* Ottawa: CRTC. 19 May.

—. 1999. *Report on new media.* Ottawa: CRTC. 17 May.

—. 2000. *Decision CRTC 2000-747. Transfer of effective control of CTV Inc. to BCE. Inc.* Ottawa: CRTC. 7 December.

—. 2001. *Broadcasting policy monitoring report.* Ottawa: CRTC. 29 October.

—. 2002a. *The CRTC's mandate.* Ottawa: CRTC. <www.crtc.gc.ca/eng/about.htm#mandate> (May 2002).

—. 2002b. *Telecommunications proceeding.* Ottawa: CRTC. <www.crtc.gc.ca/eng/publicpar_3.htm> (May 2002).

—. 2003. *Broadcasting policy monitoring report.* Ottawa: CRTC. 1 September.

Cultural Industries Sectoral Advisory Group on International Trade. 1999. *Canadian culture in a global world: New strategies for culture and trade.* Ottawa: Department of Foreign Affairs and International Trade.

Culver, Keith. 2003. The future of e-democracy: Lessons from Canada. *Open Democracy.* 13 November. <www.opendemocracy.net> (2 April 2004).

Dahlberg, Lincoln. 2001. The Internet and democratic discourse: Exploring the prospects of online deliberative forums extending the public sphere. *Information, Communication, and Society* 4 (4): 615-33.

—. 2003. The corporate transformation(s) of the online public sphere. Paper presented at the Association of Internet Researchers Annual Conference. Toronto, ON, 21 October.

Deibert, Ronald J. 1997. *Parchment, printing and hypermedia: Communication in world order transformation.* New York: Columbia University Press.

—. 2002a. Civil society activism on the World Wide Web: The case of the anti-MAI lobby. In *Street protests and fantasy parks: globalization, culture and the state,* ed. David R. Cameron and Janice Gross Stein, 88-108. Vancouver: UBC Press.

—. 2002b. The Internet and the "borderless" world. *ISUMA: Canadian Journal of Policy Research* 3(1): 113-19.

Department of Indian and Northern Affairs. 2002. *Report on Aboriginal community connectivity infrastructure.* Ottawa: Department of Indian and Northern Affairs. <www.aboriginalcanada.gc.ca/abdt/interface/cac_stats.nsf/engdoc/1.html> (12 May 2004).

DeRabbie, Doug. 1998. Electronic campaigning: Technology and the 1997 Canadian federal election. Paper presented to the Canadian Political Science Association annual general meeting, University of Ottawa, 31 May-2 June.

DFAIT (Department of Foreign Affairs and International Trade). 2003. *A dialogue on foreign policy: Report to Canadians.* Ottawa: DFAIT.

DOC (Department of Communications). Telecommission Directing Committee. 1971. *Instant world: A report on telecommunications in Canada.* Ottawa: Information Canada.

Dorland, Michael, ed. 1996. *The cultural industries in Canada.* Toronto: Lorimer.

Dowding, Martin. 2001. National information infrastructure development in Canada and the U.S.: (Re)defining universal service and universal access in the age of techno-economic convergence. PhD diss., Faculty of Information Studies, University of Toronto.

—. 2002. Universal access in IHAC and NIIAC: Transformed narrative and meaning in information policy. In *Citizenship and participation in the information age,* ed. Manjunath Pendakur and Roma Harris, 211-18. Aurora, ON: Garamond.

Dreyfus, Hubert. 2001. *On the Internet.* London: Routledge.

Dryburgh, Heather. 2001. Changing our ways: Why and how Canadians use the Internet. Ottawa: Statistics Canada. Cat. no. 56F0006XIE. <www.statcan.ca> (7 July 2001).

Dyer-Witheford, Nick. 1999. *Cyber-Marx: Cycles and circuits of struggle in high-technology capitalism.* Urbana: University of Illinois Press.

Eubanks, Virginia. 2000. Paradigms and perversions: A woman's place in cyberspace. *Computer Professionals for Social Responsibility Newsletter* 18(1). <www.cpsr.org/publications> (3 April 2001).

Feenberg, Andrew. 1999. *Questioning technology.* London: Routledge.

Franklin, Ursula. 1999. *The real world of technology.* Toronto: Anansi.

Fudge, Judy, Eric Tucker, and Leah Vosko. 2002. *The legal concept of employment: Marginalizing workers.* Ottawa: Law Commission of Canada.

Gerlach, Neil, and Sheryl Hamilton. 2002. Virtually civil: Studio XX, feminist voices and digital technology in Canadian civil society. In *Civic discourse and cultural politics in Canada: A cacophony of voices,* ed. Sherry Ferguson and Leslie Shade, 201-15. Westport, CT: Ablex.

Gibbins, Roger. 2000. Federalism in a digital world. *Canadian Journal of Political Science* 33(4): 667-90.

Gidengil, Elizabeth, André Blais, Neil Nevitte, and Richard Nadeau. 2004. *Citizens.* Canadian Democratic Audit. Vancouver: UBC Press.

Gitlin, Todd. 1998. Public sphere or public sphericules? In *Media, ritual, and identity,* ed. Tamar Liebes and James Curran, 168-74. London: Routledge.

Grant, George. 1986. *Technology and justice.* Toronto: Anansi.

—. 1998. *The George Grant reader,* ed. W. Christian and S. Grant. Toronto: University of Toronto Press.

Guérin, Daniel, and Asifa Akbar. 2003. Electronic voting methods: experiments and lessons. *Electoral Insight* 5(1): 26-30.

Gurstein, Penny. 2001. *Wired to the world, chained to the home: Telework in daily life.* Vancouver: UBC Press.

Gutstein, Donald. 1999. *E.con: How the Internet undermines democracy.* Toronto: Stoddart.

Habermas, Jürgen. 1991. *The structural transformation of the public sphere,* trans. T. Burger and F. Lawrence. Cambridge, MA: MIT Press.

—. 2001. The public sphere: An encyclopedia article. In *Media and cultural studies,* ed. M.G. Durham and D. Kellner, 102-7. Oxford: Blackwell.

Hannigan, John. 2001. Canadian media ownership and control in the age of the Internet and global megamedia empires. In *Communications in Canadian society,* 5th ed., ed. Craig McKie and Benjamin Singer, 245-55. Toronto: Thompson Publishing.

—. 2002. The global entertainment economy. In *Street protests and fantasy parks: Globalization, culture and the state,* ed. David R. Cameron and Janice Gross Stein, 20-48. Vancouver: UBC Press.

Herman, Edward S., and Robert W. McChesney. 1997. *Global media: The new missionaries of global capitalism.* London: Cassell.

Hillwatch E-services. 2004. Political web sites: Strategic assets or virtual lawn signs? Ottawa: Hillwatch Inc. <www.hillwatch.com> (15 July 2004).

House of Commons. 1932a. *Debates.* R.B. Bennett.

—. Special Committee on Radio Broadcasting. 1932b. *Minutes.* Graham Spry.

Hurrell, Christie. 2004. Shaping policy discourse in the public sphere: Civility and the design of a public sphere for online dialogue. Paper presented at the conference Communication and Democracy: Technology and Citizen Engagement, University of New Brunswick, Fredericton, 5 August.

IHAC (Information Highway Advisory Council). 1995. *Connection, community, content: The challenge of the information highway.* Final report of the Information Highway Advisory Council. Ottawa: Minister of Supply and Services.

—. 1997. *Preparing Canada for a digital world.* Final report of the Information Highway Advisory Council. Ottawa: Minister of Supply and Services.

Industry Canada. 1996. *Building the information society: Moving Canada into the 21st century.* Ottawa: Minister of Supply and Services. Cat. no. C2-302/1996.

—. 2001. *Achieving excellence: Canada's innovation strategy.* Ottawa: Industry Canada. Cat. no. C2-596/2001.

—. 2002. Mandate. <www.ic.gc.ca> (May 2002).

Innis, Harold. 1995. *The bias of communication.* Toronto: University of Toronto Press.

ITAC (Information Technology Association of Canada). 2001. Toward a culture of innovation. <www.itac.ca> (20 May 2004).

Jay's Leftist and Progressive Internet Resources Directory. 2004. Canada leftist and "progressive" links. <www.neravt.com/left/directory/countries/canadian.htm> (22 April 2004).

Johnson, Laura C. 2002. *The co-workplace: Teleworking in the neighbourhood.* Vancouver: UBC Press.

Johnston, Richard. 2000. Canadian elections at the millennium. *Choices* 6(6): 4-36.

—. 2001. A conservative case for electoral reform. *Policy Option/Options Politique* 22(6): 7-14.

Juniper, David Kim. 2002. The Moccasin Telegraph goes digital: First Nations and political usage of the Internet. In *Civic discourse and cultural politics in Canada: A cacophony of voices,* ed. Sherry Ferguson and Leslie Shade, 142-51. Westport, CT: Ablex.

Karim, Karim H. 2002. Globalization, communication and diaspora. In *Mediascapes: New patterns in Canadian communication,* ed. Paul Attallah and Leslie Regan Shade, 272-94. Scarborough, ON: Nelson.

Kellner, Douglas. 1997. Intellectuals, the new public spheres and technopolitics. *New Political Science* 41-2(Fall): 169-88.

Kernaghan, Kenneth, Nancy Riehle, and James Lo. 2003. Politicians' use of ICTs: A survey of federal parliamentarians. *Crossing Boundaries Project.* Ottawa: Centre for Collaborative Government.

Kidd, Dorothy. 2003. Indymedia.org: a new communications commons. In *Cyber-activism: Online activism in theory and practice,* ed. Martha McCaughey and Michael Ayers, 47-69. New York: Routledge.

Kingsley, Jean-Pierre. 2000. Technology in the electoral process. *Electoral Insight* 2(1): 1.

Kippen, Grant. 2000. *The use of new information technologies by a political party: A case study of the Liberal party in the 1993 and 1997 federal elections.* Vancouver: SFU-UBC Centre for the Study of Business and Government.

Klein, Naomi. 2000. *No logo: Taking aim at the brand bullies.* Toronto: Vintage.

KPMG. 1998. *Technology and the voting process: Final report prepared for Elections Canada.* Ottawa: Elections Canada.

Kroker, Arthur. 1984. *Technology and the Canadian mind: Innis/McLuhan/Grant.* Montreal: New World Perspectives.

Lenihan, Donald. 2002a. *E-government, federalism and democracy: The new governance.* Crossing Boundaries Project: Changing Government, vol. 9. Ottawa: Centre for Collaborative Government.

—. 2002b. *E-government: The municipal experience.* Crossing Boundaries Project: Changing Government, vol. 8. Ottawa: Centre for Collaborative Government.

—. 2002c. *Re-aligning governance: From e-government to e-democracy.* Crossing Boundaries Project: Changing Government, vol. 6. Ottawa: Centre for Collaborative Government.

Lessig, Lawrence. 1999. *Code: And other laws of cyberspace.* New York: Basic Books.

Longford, Graham. 2002. Net gain? E-government in Canada, the U.S. and the U.K.: Dilemmas of public service, citizenship and democracy in the digital age. Paper presented at the forty-third annual convention of the International Studies Association, New Orleans, 24 March.

Looker, E. Dianne, and Victor Thiessen. 2003. *The digital divide in Canadian schools: Factors affecting student access to and use of information technology.* Ottawa: Statistics Canada. Cat. no. 81-597-XIE.

Lorimer, Rowland, and Mike Gasher. 2001. *Mass communication in Canada,* 4th ed. Don Mills, ON: Oxford University Press Canada.

Lyon, David. 1999. *Surveillance society: Monitoring everyday life.* Buckingham: Open University Press.

—. 2003. *Surveillance after September 11.* Cambridge: Polity.

McBride, Stephen. 2003. Quiet constitutionalism in Canada: The international political economy of domestic institutional change. *Canadian Journal of Political Science* 36(2): 251-74.

McCaughey, Martha, and Michael Ayers, eds. 2003. *Cyberactivism: Online activism in theory and practice.* London: Routledge.

McChesney, Robert. 1999. *Rich media, poor democracy: Communication politics in dubious times.* Urbana: University of Illinois Press.

McDowell, Stephen, and Cheryl Buchwald. 1997. Consultation on communications policies: Public interest groups and the IHAC. Information Policy Research Program working paper no. 4, Faculty of Information Studies, University of Toronto. <www.fis.utoronto.ca/research/iprp/publications> (9 December 2004).

Mackinnon, Mary Pat. 2004. Public dialogue and other tools for citizen engagement. CPRN Public Involvement Network. Ottawa: Canadian Policy Research Networks. <www.cprn.org> (12 April 2004).

McQuillan, Blair. 2003. Engagement can be scary, DG says. *itWorld Canada.com,* 2 April. <www.itworldcanada.com> (10 April 2004).

Magder, Ted. 1996. Film and video production. In *The cultural industries in Canada: Problems, policies and prospects,* ed. Michael Dorland, 145-77. Toronto: Lorimer.

Malloy, Jonathan. 2003. To better serve Canadians: How technology is changing the relationship between members of Parliament and public servants. New Directions no. 9. Ottawa: Institute of Public Administration of Canada.

May, James H. 1998. Information technology for indigenous peoples: The North American experience. In *Digital democracy: Policy and politics in the wired world,* ed. C. Alexander and L. Pal, 220-37. Don Mills, ON: Oxford University Press Canada.

Meisel, John. 1985. The decline of party in Canada. In *Party politics in Canada,* 5th ed., ed. H. Thorburn, 98-114. Scarborough, ON: Prentice-Hall.

—. 1986. Escaping extinction: Cultural defence of an undefended border. In *Southern exposure: Canadian perspectives on the United States,* ed. David Flaherty and William McKercher, 152-68. Toronto: McGraw-Hill Ryerson.

Meisel, John, and Matthew Mendelsohn. 2001. Meteor? Phoenix? Chameleon? The decline and transformation of party in Canada. In *Party politics in Canada,* 8th ed., ed. H. Thorburn and A. Whitehorn, 163-78. Toronto: Pearson.

Menzies, Heather. 1996. *Whose brave new world? The information highway and the new economy.* Toronto: Between the Lines.

—. 1997. Telework, shadow work: The privatization of work in the new digital economy. *Studies in Political Economy* 53: 103-23.

Moll, Marita, and Leslie Regan Shade. 2004a. Vision impossible? The World Summit on the Information Society. In *Seeking convergence in policy and prac-*

*tice: Communications in the public interest.* Vol. 2., ed. M. Moll and L.R. Shade, 45-80. Ottawa: Canadian Centre for Policy Alternatives.

—, eds. 2001. *E-commerce vs. e-commons: Communications in the public interest.* Ottawa: Canadian Centre for Policy Alternatives.

—. 2004b. *Seeking convergence in policy and practice: Communications in the public interest.* Vol. 2. Ottawa: Canadian Centre for Policy Alternatives.

Mosco, Vincent. 1993. Free trade in communication: Building a world business order. In *Beyond national sovereignty: International communication in the 1990s.* ed. Kaarle Nordenstreng and Herbert I. Schiller, 193-209. Norwood, NJ: Ablex.

—. 2004. *The digital sublime.* Cambridge, MA: MIT Press.

Mosco, Vincent, and Dan Schiller, eds. 2001. *Continental order? Integrating North America for cybercapitalism.* Lanham, MD: Rowman and Littlefield.

Murdock, Graham, and Peter Golding. 2001. Digital possibilities, market realities: The contradictions of communications convergence. In *A world of contradictions: Socialist register 2002,* ed. L. Panitch and C. Leys, 110-29. London: Merlin Press.

NBTF (National Broadband Task Force). 2001. *The new national dream: Networking the nation for broadband access.* Ottawa: Minister of Supply and Services.

Nolen, Stephanie. 2000. Can the Inuit keep their voice? *Globe and Mail,* 25 July, R1.

Norris, Pippa. 2001. *Digital divide: Civic engagement, information poverty, and the Internet worldwide.* London: Cambridge University Press.

—. 2002. Revolution, what revolution? The Internet and the U.S. elections, 1992-2000. In *Governance.com: Democracy in the information age,* ed. Elaine Ciulla Kamarck and Joseph S. Nye Jr, 59-80. Washington, DC: Brookings Institution.

Office of the Auditor General of Canada. 2003. *Report of the Auditor General of Canada to the House of Commons.* Ottawa: Minister of Public Works and Government Services.

Paré, Daniel. 2004. The digital divide: Why the "the" is misleading. In *Human rights in the digital age,* ed. A. Murray and M. Klang, 85-97. London: Cavendish.

Patriquin, Martin. 2003. The charm offensive. *Xpress,* 24-30 July, 8.

Pendakur, Manjunath, and Roma Harris, eds. 2002. *Citizenship and participation in the information age.* Aurora, ON: Garamond.

Phillips, Susan D., and Michael Orsini. 2002. *Mapping the links: Citizen involvement in policy processes.* CPRN Discussion Paper no. F21. Ottawa: Canadian Policy Research Networks.

Poster, Mark. 2001. *What's the matter with the Internet?* Minneapolis: University of Minnesota Press.

Potter, Evan. 2002. Anarchy makes a comeback. In *Civic discourse and cultural politics in Canada: A cacophony of voices,* ed. Sherry Ferguson and Leslie Shade, 91-108. Westport, CT: Ablex.

Preyra, Leonard. 2001. From conventions to closed primaries? New politics and recent changes in national party leadership selection in Canada. In *Party politics in Canada,* 8th ed., ed. H. Thorburn and A. Whitehorn, 443-59. Toronto: Pearson.

Privy Council Office. 2002. *Consulting and engaging Canadians: Guidelines for online consultation and engagement.* Ottawa: Privy Council Office.

Public Works and Government Services. 2004. *Government on-line 2004.* Ottawa: Minister of Public Works and Government Services.

Raboy, Marc. 1990. *Missed opportunities: The story of Canada's broadcasting policy.* Montreal: McGill-Queen's University Press.

—. 1995. The role of public consultation in shaping the Canadian broadcasting system. *Canadian Journal of Political Science* 28(2): 455-77.

—. 1997. Cultural sovereignty, public participation, and democratization of the public sphere. In *National information infrastructure initiatives: Vision and policy design,* ed. B. Kahin and E. Wilson, 190-216. Cambridge, MA: MIT Press.

—. 2001. Cultural policy in the knowledge society. In *Vital links for a knowledge culture: Public access to new information and communication technologies,* ed. L. Jeffrey and I. Nayman, 141-58. Strasbourg: Council of Europe.

—. 2002. Communication and globalization — A challenge for public policy. In *Street protests and fantasy parks: Globalization, culture and the state,* ed. David R. Cameron and Janice Gross Stein, 109-40. Vancouver: UBC Press.

Raboy, Marc, Ivan Bernier, Florian Sauvageau, and Dave Atkinson. 1994. Cultural development and the open economy: A democratic issue and a challenge to public policy. *Canadian Journal of Communication* 19(3-4): 1-25.

Reddick, Andrew, and Christian Boucher. 2002. Tracking the dual digital divide. Ottawa: Ekos Research. <www.ekos.com/media/files/dualdigitaldivide.pdf> (8 May 2004).

Rideout, Vanda. 2000. Public access to the Internet and the Canadian digital divide. *Canadian Journal of Information and Library Science* 25(2-3): 1-21.

Rose, Jonathan, and Tamara Small. 2004. Engaging citizens or engaging the press? An examination of the 2004 Conservative leadership campaign on-line. Paper presented at the conference Communication and Democracy: Technology and Citizen Engagement, University of New Brunswick, Fredericton, 5 August.

Royal Commission on Radio Broadcasting. 1929. *Report.* Ottawa: King's Printer.

Salter, Lee. 2003. Democracy, new social movements and the Internet: A

Habermasian analysis. In *Cyberactivism: Online activism in theory and practice,* ed. Martha McCaughey and Michael Ayers, 117-44. New York: Routledge.

Sassen, Saskia. 2002. Mediating practices: women with/in cyberspace. In *Living with cyberspace: technology and society in the 21st century,* ed. J. Armitage and J. Roberts, 109-19. London: Continuum.

Schiller, Dan. 1999. *Digital capitalism: Networking the global market system.* Cambridge, MA: MIT Press.

Schiller, Dan, and Vincent Mosco. 2001. Introduction: Integrating a continent for a transnational world. In *Continental order? Integrating North America for cybercapitalism,* ed. Vincent Mosco and Dan Schiller, 1-34. Lanham, MD: Rowman and Littlefield.

Sennett, Richard. 1998. *The corrosion of character: The personal consequences of work in the new capitalism.* New York: Norton.

Shade, Leslie Regan. 1999. Net gains: Does access equal equity? *Journal of Information Technology Impact* 1: 23-39.

—. 2002. *Gender and community in the social construction of the Internet.* New York: Peter Lang.

Smith, Peter, and Elizabeth Smythe. 2001. Globalization, citizenship and technology: The Multilateral Agreement on Investment meets the Internet. In *Culture and politics in the information age: A new politics?* ed. Frank Webster, 183-206. London: Routledge.

Smythe, Dallas W. 1981. *Dependency road: Communications, capitalism, consciousness and Canada.* Norwood, NJ: Ablex.

Spender, Dale. 1996. *Nattering on the net: Women, power and cyberspace.* Toronto: Garamond.

Spry, Graham. 1931. The Canadian radio situation. In *Education on the air,* ed. J.H. Maclatchy, 47-60. Columbus: Ohio State University Press.

Standing Committee on Canadian Heritage. 2003. *Our cultural sovereignty: The second century of Canadian broadcasting.* Ottawa: House of Commons.

Statistics Canada. 2003. Household Internet use survey. *The Daily.* 18 September. <www.statcan.ca> (12 November 2003).

—. 2004. 2001 Aboriginal peoples survey: Community profiles. Ottawa: Statistics Canada. <http://www12.statcan.ca/english/profil01aps/home.cfm> (25 April 2004).

Stewart Millar, Melanie. 1998. *Cracking the gender code: Who rules the wired world?* Toronto: Second Story.

Straw, Will. 1996. Sound recording. In *The cultural industries in Canada: Problems, policies and prospects,* ed. Michael Dorland, 95-117. Toronto: Lorimer.

Sunstein, Cass. 2001. *Republic.com.* Princeton: Princeton University Press.

Taras, David. 2001. *Power and betrayal in the Canadian media.* Peterborough, ON: Broadview.

Treasury Board Secretariat. 1993. *Blueprint for renewing government services using information technology.* Ottawa: Minister of Supply and Services.

—. 2003. *Government on-line 2003.* Ottawa: Minister of Public Works and Government Services.

Turner, Michael. 2003. Towards a centre of expertise. Online Consultation Technologies Centre of Expertise, speech, 8 September. Ottawa: Public Works and Government Services Canada. <www.pwgsc.gc.ca/econsultation/text/mturner_speech.html> (14 April 2004).

UNESCO. 1980. *Many voices, one world.* Report of the International Commission for the Study of Communication Problems. London: Kogan Page.

Valaskakis, Gail. 2002. Remapping the Canadian north: Nunavut, communications and Inuit participatory development. In *Citizenship and participation in the information age,* ed. Manjunath Pendakur and Roma Harris, 400-14. Aurora, ON: Garamond.

Van Rooy, Allison. 2000. *Electronic consultation and engagement: Lessons for Canada from international experience.* Ottawa: Privy Council Office, Policy and Research, Intergovernmental Affairs.

Vosko, Leah. 2000. *Temporary work: The gendered rise of a precarious employment relationship.* Toronto: University of Toronto.

Vosko, Leah, Nancy Zukewich, and Cynthia Cranford. 2003. Precarious jobs: A new typology of employment. *Perspectives on Labour and Income* 15(4): 16-26. Statistics Canada. Cat. no. 75-001-X1E.

Walters, Gregory. 2001. Information highway policy, e-commerce and work. In *E-commerce vs. e-commons: Communications in the public interest,* ed. Martin Moll and Leslie Regan Shade, 69-74 Ottawa: Canadian Centre for Policy Alternatives.

Warschauer, Mark. 2002. *Technology and social inclusion: Rethinking the digital divide.* Cambridge, MA: MIT Press.

Webster, Frank, ed. 2001. *Culture and politics in the information age: A new politics?* London: Routledge.

Whitaker, Reginald. 1999. *The end of privacy: How total surveillance is becoming a reality.* New York: The New Press.

White, Graham. 2000. And now for something completely northern: Institutions of governance in the territorial north. *Journal of Canadian Studies* 35(4): 80-99.

Wilhelm, Anthony. 2000. *Democracy in the digital age.* London: Routledge.

Williams, Raymond. 1983. *Culture and society 1780-1950.* New York: Columbia University Press.

Winner, Langdon. 1986. *The whale and the reactor: A search for limits in an age of high technology.* Chicago: University of Chicago Press.

—. 1995. Citizen virtues in a technological order. In *Technology and the politics of knowledge,* ed. A. Feenberg and A. Hannay, 65-84. Bloomington: University of Indiana Press.

Winseck, Dwayne. 1998. *Reconvergence: A political economy of telecommunications in Canada.* Cresskil, NJ: Hampton Press.

—. 2002a. Illusions of perfect information and fantasies of control in the information society. In *Citizenship and participation in the information age,* ed. Manjunath Pendakur and Roma Harris, 33-55. Aurora, ON: Garamond.

—. 2002b. Netscapes of power: Convergence, consolidation and power in the Canadian mediascape. *Media, Culture and Society* 24(6): 795-819.

Young, Lisa, and Joanna Everitt. 2004. *Advocacy groups.* Canadian Democratic Audit. Vancouver: UBC Press.

# Index

A master index to all volumes in the Canadian Democratic Audit series can be found at www.ubcpress.ca/readingroom/audit/index.